THE

FOSSIL FLORA

OF

GREAT BRITAIN;

OR,

FIGURES AND DESCRIPTIONS

OF THE

VEGETABLE REMAINS FOUND IN A FOSSIL STATE

IN THIS COUNTRY.

BY

JOHN LINDLEY, Ph. D. F.R.S. &c.

PROFESSOR OF BOTANY IN UNIVERSITY COLLEGE, LONDON;

AND

WILLIAM HUTTON, F.G.S. &c.

"Avant de donner un libre cours à notre imagination, il est essentiel de rassembler un plus grand nombre de faits incontestables, dont les conséquences puissent se déduire d'elles-mêmes."—*Sternberg*.

VOLUME III.

LONDON:

JAMES RIDGWAY AND SONS, PICCADILLY.

MDCCCXXXVII.

THE

FOSSIL FLORA.

NOTE

UPON THE VALUE OF NUMERICAL PROPORTIONS IN THE ANCIENT FLORA OF THE WORLD, WITH REFERENCE TO A DETERMINATION OF CLIMATE.

BY PROFESSOR LINDLEY.

“ Si nous comparons cette flore ancienne avec les flores des diverses régions du globe, sous le point de vue de la proportion numérique des espèces des différentes classes, nous n'en trouverons aucune qui lui soit complétement analogue ; mais nous verrons que plus ces flores appartiennent à des espaces de terre plus circonscrits au milieu d' étendues d'eau plus vastes, c'est à dire à des îles plus petites et plus éloignées des continens, et plus elles se rapprochent par la proportion des diverses familles de ce que nous connaissons dans les terrains houillers. Suivant l'observation faite en premier, je crois, par M. R. Brown, et qui a été développée depuis par M. d'Urville, les Fougères et les Lycopodes paroissent soumises à deux influences différentes, qui déterminent les nombres des espèces de ces familles par rapport au nombre total des végétaux phanérogames : la température est une de ces causes; l'influence de l'air humide et de la température uniforme da la mer, paraît être l'autre. Il en résulte que dans les localités également favorisées sous le rapport de ces dernières circonstances, ces plantes sont plus fréquentes dans la zone équatoriale que dans les zones plus froides; mais que sous la même zone elles sont beaucoup plus abondantes dans les îles que sur les continens. Nous pourrions citer de nombreux exemples à l'appui de cette proposition, mais ce n'en est pas ici le lieu; nous dirons seulement que dans les parties les plus favorables au développement de ces plantes sur le continent de l'Europe tempérée, leur rapport aux phanérogames est comme 1:40, tandis que dans les mêmes circonstances, dans les régions continentales, entre les tropiques, M. R. Brown admet que ce rapport est comme 1:20, et dans les cas moins favorables comme 1:26.

" Sous la même latitude cette proportion devient bien plus grande dans les îles: ainsi, dans les Antilles le rapport des Fougères aux plantes Phanérogames paraît être à peu près comme 1:10, au lieu de 1:20, qui est celui des parties les plus favorisées du continent Américain; dans les îles de la mer du Sud, ce rapport, au lieu d'être 1:26, comme dans le continent de l'Inde et de la Nouvelle Hollande tropicale, devient 1:4, ou 1:3; à Sainte-Hélène et à Tristan d' Acugna la proportion de ces végétaux est comme 2:3; enfin, à l'île de l'Ascension, en ne considérant que les plantes évidemment indigènes, il paraît y avoir égalité entre les plantes Phanérogames et les Cryptogames vasculaires.

" On conçoit donc que, si des îles analogues à celles que nous venons de citer, existaient seules sur la surface de notre globe au milieu d'une vaste mer, où elles ne formeraient que des sortes de points épars, la proportion de Fougères serait probablement encore plus grande, et, au lieu de l'égalité des deux grands groupes de végétaux que nous comparons, nous pourrions voir les Cryptogames vasculaires l'emporter de beaucoup sur les Phanérogames; c'est ce qui a lieu dans le terrain houiller, et ces considérations de géographie botanique doivent déjà nous porter à penser que les végétaux qui ont donné naissance à ces dépôts, croissaient sur des archipels d'îles peu étendues. La disposition des terrains houillers par lignes interrompues, qu'on a appelés des bassins et comparés à des successions de lacs ou à des vallées, est au moins aussi analogue à la disposition la plus frequente des îles qui, représentant les çrêtes de chaînes de montagnes sous-marines, sont généralement placées en séries ; enfin le morcellement du terrain houiller, et au contraire la vaste étendue et la continuité des terrains de calcaire de transition, qu'on peut considérer comme les dépôts formés dans la mer qui environnait ces îles, nous semblent confirmer cette hypothèse." *Adolphe Brongniart, Prodr. p.* 181.

Such were the opinions entertained by Monsieur Adolphe Brongniart in the year 1828, and such are probably the opinions of many Geologists at the present day; for there certainly has as yet been nothing done or discovered to call their soundness in question.

It however always appeared to me very doubtful whether such data as we possessed concerning the Flora of the Coal Measures could be considered of a nature sufficiently precise to justify Geologists in entering into such calculations, in which, for them to be of any value whatever, a full knowledge of all facts is obviously indispensable. It was, moreover, perfectly clear that the numerical proportion borne by Ferns to other plants was rapidly diminishing as the examination of the vegetable remains of the Coal Measures became more carefully conducted. The very remarkable fact that Ferns are scarcely ever met with in fructification in a fossil state was also a circumstance upon which no light was thrown by the theory of a high temperature, and damp insular atmosphere.

Taking all these into consideration, along with the constant state of disintegration of vegetable remains—a disintegration unquestionably not the result of drifting—I was led to suspect that possibly the total absence of certain kinds of plants, the as constant presence of others, and several other points of a like nature, might be accounted for by a difference in the capability of one plant beyond another of resisting the action of water.

Accordingly, on the 21st of March, 1833, I filled a large iron tank with water, and immersed in it 177 specimens of various plants, belonging to all the more remarkable natural orders, taking care in particular to include representatives of all those which are either constantly present in the Coal Measures, or as universally absent. The vessel was placed in the open air, left uncovered, and left untouched, with the exception of filling up the water as it evaporated, till the 22nd of April, 1835, that is, for rather more than two years. At the end of that time what remained was examined with the results stated in the following list; in which it is to be observed that where no observation is added to the name of a plant, no trace whatever of that species could be found.

ACOTYLEDONES.

Fungi.	*Result.*
1 Boletus suberosa.	A black shapeless mass.
2 ——— versicolor.	Ditto.
3 ——— sp.	Ditto.
Lichenes.	
4 Peltidea canina.	
5 Parmelia saxatilis.	
6 Thelotrema pertusum.	
Hepaticæ.	
7 Marchantia polymorpha.	
Musci.	
8 Hypnum striatum.	
9 ——— sericeum.	
10 Dicranum purpureum.	
11 ——— scoparium.	
12 Bryum undulatum.	
13 Polytrichum commune.	
Filices.	
14 Aspidium Filix mas.	In good condition, but the fructification rotted off.
15 ——— aculeatum.	Ditto ditto.
16 Pteris aquilina, (dead leaves.)	Much broken and decayed, scarcely to be recognized.
17 Scolopendrium vulgare.	Good condition, no fructification.
18 Polypodium vulgare.	Recognizable but decayed.
19 ——— cambricum. (dead leaves.)	
Lycopodiaceæ.	
20 Lycopodium Phlegmaria, (dried.)	Good condition.
Equisetaceæ.	
21 Equisetum hiemale.	
22 ——— variegatum.	

DICOTYLEDONES.

Cycadeæ.	*Result.*
1 Zamia horrida.	Pinnæ quite perfect; but they had separated from their petiole, leaving a double range of oblique narrow holes in its front.
2 —— elegans.	Quite perfect, except near the base where the pinnæ had dropped off, leaving holes as in the last instance.
Coniferæ.	
3 Thuja orientalis.	Decayed, but recognizable.
4 —— occidentalis.	Ditto, ditto.
5 Juniperus virginiana.	Many leaves fallen off; much decayed.
6 —— Sabina.	Good condition.
7 Pinus Pinea.	Leaves in good condition; but mostly fallen off.
8 —— halepensis.	Ditto, ditto.
9 Abies balsamea.	Ditto, ditto.
10 —— canadensis.	Ditto, ditto.
11 —— rubra.	Very perfect; leaves still adhering.
12 —— Webbiana.	
13 —— Cedrus.	All the leaves fallen off; bad condition.
14 Cunninghamia lanceolata.	Nearly perfect.
15 Araucaria imbricata.	Quite perfect.
16 —— excelsa.	Branch only left, leaves lost; not distinguishable.
17 —— Cunninghami.	(same as 16)
18 Taxus baccata.	
Amentaceæ.	
19 Fagus sylvatica, (dry leaves.)	
20 Carpinus Betulus, (ditto.)	
21 Platanus orientalis, (ditto.)	

No.	Species	*Results.*
22	Quercus Ilex.	Good condition. (22–24)
23	——— suber.	
24	——— austriaca.	
25	——— pedunculata, (dry leaves.)	
26	——— Cerris, (ditto.)	Recognizable.
	Miscellaneous Apetalæ.	
27	Buxus communis.	
28	——— balearica.	Good condition.
29	Croton variegatum.	
30	Myrica cerifera.	Tolerably perfect.
31	Rumex Patientia.	
32	Coccoloba uvifera.	Perfect.
33	Laurus fœtens.	
34	——— camphora.	Reduced to a skeleton.
35	Casuarina equisetifolia.	Nearly perfect.
36	Dryandra speciosa.	Tolerably perfect.
37	Ficus Brassii.	
38	——— elastica.	Nearly perfect.
	Polypetalæ.	
1	Magnolia grandiflora.	
2	Berberis glumacea.	
3	——— repens.	
4	——— Aquifolium.	
5	——— fascicularis.	
6	Hypericum calycinum.	
7	Photinia serrulata.	
8	Eucalyptus pulverulenta.	
9	Myrtus communis.	
10	Mimosa scandens.	
11	Eugenia macrocarpa.	
12	Schinus Litri.	
13	Ligusticum Levisticum.	
14	Gastonia palmata.	
15	Sanguisorba officinalis.	
16	Fragaria virginiana.	
17	Eriobotrya japonica.	
18	Prunus Lauro-cerasus.	

		Results.
19	Prunus lusitanica.	
20	Acacia verticillata.	
21	Bauhinia racemosa.	
22	Spartium junceum.	
23	Ceratonia Siliqua.	
24	Spartium scoparium.	
25	——— multiflorum.	
26	Brassica caulorapa.	
27	Cereus speciosus.	
28	——— brasiliensis.	
29	Saxifraga crassifolia.	
30	Tellima grandiflora.	
31	Oxalis acetosella.	
32	Aristotelia Maqui.	
33	Dodonæa triquetra.	
34	Echeveria gibbiflora.	
35	Cotyledon sp.	
36	Francoa appendiculata.	
37	Ribes punctatum.	
38	Passiflora racemosa.	
39	Hibiscus liliiflorus.	
40	Reevesia sinensis.	In tolerable condition.
41	Pterospermum acerifolium.	A perfect skeleton remaining.
42	Astrapæa Wallichii.	
43	Banisteria chrysophylla.	
44	Helleborus odorus.	
45	Hedera Helix.	
	Monopetalæ.	
1	Ilex Aquifolium.	
2	—— balearica.	
3	Phillyrea obliqua.	
4	——— latifolia.	
5	——— angustifolia.	
6	Jasminum revolutum.	
7	Olea europæa.	
8	Rhododendron ponticum.	Leaf reduced to a skeleton.
9	——— azaleoides.	
10	Kalmia latifolia.	

	Results.
11 Arbutus Unedo.	Much decayed.
12 Andromeda calyculata.	
13 ———— speciosa.	
14 ———— pulverulenta.	
15 Gaultheria Shallon.	Good condition.
16 Ledum latifolium.	
17 Caprifolium implexum.	
18 Viburnum sinense.	
19 ———— Tinus.	
20 Aucuba japonica.	
21 Bignonia capreolata.	
22 Acanthus mollis.	
23 Bignonia adenophylla.	
24 Gesneria bulbosa.	
25 Gloxinia speciosa.	
26 Theophrasta Jussiæi.	Good condition.
27 Corynocarpus lævigatus.	Good condition.
28 Fagræa obovata.	
29 Brexia spinosa.	Tolerably perfect.
30 Clerodendron hastatum.	
31 Antirrhinum majus.	
32 Rosmarinus officinalis.	
33 Salvia officinalis.	
34 Phlomis ferruginea.	
35 Aster argophyllus.	
36 Sonchus arboreus.	
37 Brachyglottis repanda.	
38 Mikania Guaca.	
39 Tussilago fragrans.	
40 Cyclamen europæum.	
41 Primula sinensis.	

MONOCOTYLEDONES.

Palmæ.	
1 Phœnix dactylifera.	Good preservation.
Miscellaneous.	
2 Yucca angustifolia.	
3 ——- gloriosa.	Traces only left in the form of a thin striated blade.
4 ——- filamentosa.	

		Result.
5	Ruscus aculeatus.	Good condition.
6	——— hypoglossum.	
7	——— racemosus.	
8	Phormium tenax.	Thin film only left.
9	Peliosanthes Teta.	
10	Dichorisandra thyrsiflora.	
11	Tradescantia discolor.	
12	Dracæna terminalis.	
13	Pancratium amænum.	
14	Doryanthes excelsa.	
15	Tillandsia farinosa.	Traces with the marginal spines perfect; otherwise they could not have been recognized.
16	——— sp.	
17	——— sp.	
18	Ananassa sativa.	Good condition.
19	Iris Pseud-acorus.	
20	Renealmia nutans.	
21	Maranta zebrina.	
22	——— ramosissima.	
23	Bletia Tankervilliæ.	
24	Canna indica.	Good condition.
25	Caladium esculentum.	
26	Arum Dracunculus.	
27	Bambusa.	
28	Poa aquatica.	
29	Carex Œderi.	
30	——— pendula.	
31	Juncus conglomeratus.	

Besides these, small branches of *Elder, Oak, Horsechesnut, Plane, Sycamore, Poplar, Ash,* and *Laburnum,* were placed in the water; when examined they had all lost their bark, and could no longer be distinguished by any external character.

General Result of the preceding Experiment.

	Number of Species submitted to experiment.		Recognizable afterwards.		Not to be traced.	
ACOTYLEDONES.						
Fungi..................		3		3		0
Lichenes		3		0		3
Hepaticæ		1		0		1
Musci..................		6		0		6
Filices		6		6		0
Lycopodiacæ		1		1		0
Equisetaceæ		2		0		2
Total..	22		10		12	
DICÒTYLEDONES APETALÆ.						
Cycadeæ................		2		2		0
Coniferæ		16		13		3
Amentaceæ		8		4		4
Miscellaneous		12		7		5
Total..	38		26		12	
DICOTYLEDONES POLYPETALÆ	45		2		43	
DICOTYLEDONOUS MONOPETALÆ	41		6		35	
MONOCOTYLEDONES.						
Palmæ		1		1		0
Miscellaneous		30		11		19
Total..	31		12		19	
Total..	177		56		121	

This experiment appears to me to lead to most important conclusions. These things seem clear: firstly, that Dicotyledonous plants, in general, are unable to remain for two years in water without being totally decomposed; and that the principal part of those which do possess the power, are *Coniferæ* and *Cycadeæ*, which are exactly what we find in a Fossil state; secondly, that Monocotyledones are more capable of resisting the action of water, in particular Palms and Scitamineous plants, which are what we principally find as Fossils, but that Grasses and Sedges perish; so that we have no right to say that the earth was not originally clothed with Grasses because we no longer find their remains; thirdly, that Fungi, Mosses, and all the lowest forms of vegetation disappear, and that even Equisetum leaves no trace behind, which seems to settle the question of Calamites being an extinct form of that genus; and, finally, that Ferns have a great power of resisting water, if gathered in a green state, not one of them having disappeared during the experiment; but that the effect of immersion in water is to cause their fructification to rot away.

Hence the numerical proportion of different families of plants found in a fossil state throws no light whatever upon the ancient climate of the earth, but depends entirely upon the power which particular families may possess, by virtue of the organization of their cuticle, of resisting the action of the water wherein they floated, previously to their being finally fixed in the rocks in which they now are found.

PECOPTERIS ACUTIFOLIA.

Neuropteris acutifolia. *Murray MSS.*

Found by Dr. Murray in a new bed of vegetable remains discovered by himself, at a spot where the Sandstone enclosing the shale, passing under the Cornbrash and Kelloways rock, was denuded by the falling of the rocks at the foot of the lofty cliffs which guard Redcliffe Bay; just at the extreme of high water mark. It was accompanied as usual in these Oolitic rocks by jet and pyritous Dicotyledonous woods.

It appears to be a species of Pecopteris, distinct from any previously noticed, but is very like the following from the same locality; differing, however, in its lobes being much more pointed. It is also allied to Pecopteris tenuis, but is smaller in all its parts, and appears from the drawing to have much fewer secondary veins.

We have to thank Mr. Williamson, Jun., for the drawings and notes that accompanied them.

Fig. 1. *c.* represents a magnified view of a portion of a leaf of *Pecopteris obtusifolia* contrasted with a similar portion of the present species, *fig.* 2. *b.*

PECOPTERIS OBTUSIFOLIA.

Neuropteris obtusifolia. *Murray MSS.*

Discovered with the last by Dr. Murray.

Mr. Williamson, Jun. has communicated the following note with his drawing.

"From the thickness of the small fragment of stem remaining it has been a bipinnated frond, of rather a large size. The stem, as is usual with Ferns, is irregularly grooved or striated; the central stalk of the pinnules is straight, gradually tapering, and has the strongly marked groove along the centre: but whether the pinnules have been opposite or alternate I cannot say. They are alternate and almost an inch in length. The midrib appears very small, but the centre of the pinnule is a little prominent, as though from the pressure of a strong rib at the opposite side; the lobes are small, blunt, alternate, and extremely regular; attached by the whole of their base, which is but little broader than the apex. On each side of the midrib is a series of small sori, varying in number from 4 to 8, but generally 6; they appear as if unconnected with secondary veins, which are, however, visible in pinnules where there are no sori." See plate 157, 1. *c.*

Fig. 2. and 2. *a.* represent fragments of Pecopteris acutifolia.

SPHÆREDA PARADOXA.

" A plant occuring in the lower Shale and Sandstone at Cloughton, and which seems to throw some light upon the nature of those singular spherical bodies often seen in the shale, both there and at Gristhorpe in the upper beds, and which hitherto have been called winged seeds. The fossil under consideration, has probably been the stem or radical shoots of a Plant very analogous to, if not quite identical with the existing genus *Pilularia*, which has round sporules attached by similar very short stalks to the main root, so short indeed as nearly to appear sessile, as they do in this specimen, but in others, the capsule may be traced with a short peduncle.

" The central supporting stem appears from its its irregularity and general structure, more similar to a root than to the stem of any vegetable, especially as not any leaves can be observed in this or other specimens; and possibly it may be owing to the extreme fragility of roots that these seed vessels have been so rarely found in position.

"The structure of the capsules would seem to have consisted in an inner nucleus, protected by a very firm cortical case, which occasionally is separable, and which is an additional point of resemblance with the recent Pilularia, only that the extinct species must have been quite a giant to its present analogue.

"Now as to the theoretical deductions from the vast frequency of such plants in our Oolitic formation as Solenites, congeneric or almost identical with Isoëtes, and as Pilularia, both existing not merely in wet marshy ground, but actually in water itself, what can we infer, but that the whole of the district wherein they now occur, was a morass deeply covered in many places with water; and that *fresh* water, in which such plants grew and floated, and which along with the Lycopodia, and Ferns, some large and arborescent, others humble and delicate, were all suddenly overwhelmed by an irruption of the ocean, when the saline impregnation soon would destroy vegetable life in such plants, and the sedimentary deposits through a long series of ages, would gradually produce the slaty clay, and granular sandstone now inclosing those beautiful and interesting remains.

"Assuredly fire could not have been the agent of transmutation, because both resin and tannin have been detected still existing in the scarcely mineralized leaves of the Solenites, and of one or two species of Cyclopteris, and such vegetable principles, though not changed by the action of water, must have been decomposed by intense heat, or any combustion."

Thus far Dr. Murray concerning this remarkable production, upon which we are really unable to offer any opinion. We have inspected the specimens through the liberality of Dr. Murray; but they throw no light upon their original nature so far as we can discover.

Mr. Williamson, Jun., writes concerning the plant as follows.

"It is very coarsely and irregularly striated, and the curious lateral appendages branch off without any apparent uniformity of direction. With them, but separate and detached, are found many singular berry-like substances, and the question is, have they any connection with the plant. The stems of the smaller specimens are striated, but more finely, and the lateral branches appear to have been terminated by a kind of round or oval leaf, which is now one homogeneous mass of carbon without the least appearance of any regular veins or striæ either in the stems or leaves. The carbon is excessively thick, instead of being in thin laminæ as in most vegetable impressions."

SPHENOPTERIS CRASSA.

From the Limestone of Burdiehouse.

We know no modern Fern in which so great a disproportion exists between the pinnules and rachis as in this very curious plant. There is also the remarkable circumstance connected with it of some of the pinnules being very short and others much longer.

The general form of the latter appears to have been nearly round, with three or more deep lobes, and a somewhat doubly crenelled margin. We find, however, no regularity in the form of the pinnules nor in their size.

The stem is remarkably speckled, but whether this arises from original markings, or from minute fractures of the carbonaceous coating of the surface we are uncertain.

We have seen no trace of this species in any other rock than that of Burdiehouse.

NOTE UPON THE BURDIEHOUSE FORMATION.

BY MR. HUTTON.

In the present number we figure two more Fossils from that remarkable deposit of the remains of the Vegetable and Animal kingdoms, that occurs at Burdiehouse near Edinburgh. In the year 1831, we first visited the quarries in company with our excellent friend, Mr. Witham, in whose collection we had observed some very fine specimens from this locality, two of which are figured at plates 45 and 53, of this work, which were the first of the Fossils from Burdiehouse that were published.

It had been our intention to devote one entire portion of the Fossil Flora, to the elucidation of the Fossils of this locality, and we were aware that through our observant friends, every new plant found would be made known to us, but the variety of the species occurring, bears no comparison to the number of remains; and, notwithstanding the liberality of the Royal Society of Edinburgh, who, through the kindness of their Secretary, John Robison, Esq. placed the whole of their collection at our disposal, and also that of Dr. Hibbert, who was kind enough to lend specimens of all he thought peculiar, we are yet unable to make out satisfactorily more than a very few species.

The Fossils of Burdiehouse occur in a bed of Limestone 27 feet thick, remarkably compact, uncrystalline, and uniform throughout; the Geological position of this Limestone, is low down, probably very near the base of the Carboniferous group of rocks, it is highly inclined, dipping with its immediately associated beds, at an angle of 23° in a S.E. direction, from the Trap of the Pentland Hills, the protrusion of which has evidently thrown them into the position they now hold.

The vegetable remains occur in great profusion, and are to be found in every part of the Limestone from top to bottom, and also, but more sparingly, in the Shale beds, above and below it. There are certain well defined natural partings, or seams of stratification in the rock, which as they materially assist the working of the quarry, often expose an even surface of considerable extent, in these partings the remains of plants occur in greater abundance than any where else.

When we last visited the quarry (May 1835,) a large space was thus uncovered, which was thickly strewn over with elegant vegetable forms most perfectly preserved, the black colour of the carbonized plants, contrasting beautifully with the light lavender blue of the Limestone.—Some idea may be formed of the profusion of the remains thus exhibited, when we state that in a space of 3 feet square, we counted upwards of 40 specimens of Lepidostrobi, intermixed with Lepidophyllites almost without number, whilst scattered here and there might be observed the elegant form of Sphenopteris Affinis; so agreeable was the impression produced by the elegance of the forms, and the sober contrast of colour in the stone, that it struck us, the Calico Printer or Paper-stainer might here obtain a beautiful and certainly a novel device, for the ornament of his manufacture.

As before observed, vegetable remains occur every where throughout the Limestone, and their greater profusion in the natural partings of the rock, may be accounted for by supposing these to indicate (as they no doubt do) a period of repose—a short cessation of the calcareous deposit, whilst the parts of plants were constantly falling or being washed into the lake. There are doubtless many discoveries yet to make of vegetable forms entombed in this interesting spot, but perhaps it is somewhat unfortunate for our branch of the subject, that the brilliancy of the discoveries in Fossil Zoology, as well as the beauty and variety of the remains of animals which occur, have directed too exclusive an attention to that department. Whilst the animal remains are sought for with avidity those of plants go to the limekiln by hundreds.

Amongst vegetables the characteristic Fossils of this deposit are Lepidostrobi, Lepidophyllites, Lepidodendra, and Filicites; the rarity of Calamites, which occur but seldom, and of a diminutive size, and the almost entire absence of Stigmaria, are very striking, to those who are accustomed to view the Fossil groups usually presented by the beds of the Carboniferous formation; whilst the profusion of Lepidostrobi and Lepidophyllites of various sizes and in various states of growth, associated with the stems of Lepidodendra and those of no other plant, is an additional argument for the opinion, which has always appeared highly probable, that they are the fruit, leaves, and stem, of the same tribe of plants. Of Sigillaria, a plant which in the Flora of the Carboniferous group, generally is of so much importance, we could not observe a trace. No stems of Lepidodendra, equal in magnitude to the larger individuals found in the Coal strata and other beds of the carboniferous deposit, have yet been observed here; short portions of those of a smaller size, are met with frequently, but these are invariably turned into coal, and have lost a good deal of character, by the indistinctness of their outward form. It struck us as rather a singular circumstance, that whilst cones and leaves, and even the delicate organization of Ferns, were completely preserved, the majority of these robust stems had so little of their strongly marked character remaining. Mere carbonization does not always destroy the outward form of Fossils, and if it had in the instance of these stems, we should have expected to have found their impressions upon the Limestone matrix, as it occurs upon the Shales, and even the rough grained Sandstones, of the Coal measures. May we not suppose these to have been portions of stems (for they are mere unconnected fragments) decorticated by age and exposure, before they were deposited here? The smaller stems of Lepidodendra not unfrequently are found intimately associated with Lepidostrobi, and in some instances the cone and stem are seen in actual contact, but never in such a way as to point out with any thing like certainty, that they were parts of the same plant; perhaps, even as a collateral proof, the mere circumstance of this intimate association is not of much value, as from the abundance of Lepidostrobi, we ought to find them in connexion with every other Fossil in the deposit;—from their abundance, however, in a detached state, we may fairly infer that these cone-like bodies were easily disarticulated.

Although the vegetable remains enclosed in this bed are fragments only, yet from their size and character there is every reason to believe they have belonged to old and full grown plants, whilst from the perfect state of preservation in which the most delicate of them occur, we must suppose that the plants themselves grew in the immediate neighbourhood of the lake, and on the banks of the streams that fed it, into which portions of them were constantly falling, or that partial floods covering the land, carried off the lighter parts only, from amongst the larger stems, and deposited them where we now find their remains.

One of the many remarkable circumstances attending this bed of Limestone is, that all its organic remains proclaim it to have been produced in fresh water; in this it differs from the characters of the Calcareous beds of the Carboniferous formation generally.

Dr. Hibbert has in a luminous Memoir which he communicated to the Royal Society of Edinburgh, and which is published in their Transactions (vol. 13.), fully established the fresh water character of the Limestone of Burdiehouse, and it is to his scientific zeal, which was ably seconded by that of Mr. Robison, that we are indebted for a complete knowledge of the organic contents of this most curious deposit. To the Royal Society of Edinburgh also, as a body, the scientific world are deeply indebted, they having promptly stepped forward, at the suggestion of Dr. Hibbert, and with a power which could be commanded by no individual, rescued the Fossils from destruction and dispersion, preserving to themselves one entire set, which they most liberally lay open to all those who feel an interest in them.

The Burdiehouse bed, after a considerable interval filled with alternations of Sandstone, Shale, and Coal, is succeeded by another thick stratum of Limestone, having the usual characters of those of this formation, all its Fossils being of marine origin. We should not be surprised to find this much more generally the case than is at present supposed, as many of the Coal, Shale, and Sandstone beds, with which Limestone is associated, and which form by far the larger portion of the Carboniferous group, bear undoubted marks of their origin in fresh water. Dr. Hibbert mentions several beds of Limestone near Edinburgh, besides that of Burdiehouse, which possess this character. Mr. Murchison, also, describes one as occurring in the Coalfield near Shrewsbury, and we have ourselves had occasion to observe a belt of Limestone, almost of the same age and Geological position, as the Burdiehouse bed, which is worked along with a thin seam of coal, on the tops of the hills, west and south-west of Wooler in Northumberland; where the remains of Lepidodendra and Stigmaria are associated with those of the Cypris, and its allied genera of Entomostraca.

These alternations of salt and fresh water deposits, which are so well known in the newer formations, have led to the idea of a series of oscillatory movements, by which the surface of the earth, during the deposition of the strata, was alternately brought within the influence of the ocean and fresh water.

It has been demonstrated that very large portions of the Earth's surface have been thrust up far above the level at which they were produced, and there are many reasons for supposing that this was a gradual and not a sudden operation.

In the splendid theory of Mons. E. de Beaumont, this forcing up is supposed to have arisen, from the endeavours made by the outer cooled crust of the earth, to adjust itself to the interior mass, which was constantly losing bulk, in its passage from a state of igneous fluidity, to one of hardness—cracks being formed through which chains of mountains were protruded. Now supposing this, or any thing analogous to it, to have taken place, the gravitation of the large masses of hardened matter, between these lines of fissure or elevation, would form hollows, which would gradually deepen as the mountains rose.

If the depression of surface was great enough to bring it within the operation of the sea, and no barrier intervened, then would deposits be formed containing marine remains, layer upon layer, differing in nature according as the silicious, aluminous, or calcareous matter predominated in the waters. This would continue until the surface was so far raised as to shut out the sea, then fresh-water deposits would succeed, differing in their character, according to the nature of the detritus brought down by the operation of floods—until at length there was a surface partially dry, fitted for the growth of vegetables; these, from the abundance of moisture, and the high degree of temperature existing, would grow rapidly and to a large size, until, by successive growth and decay, their remains formed a thick layer—a sort of peat-bog—

which, in its turn, by gradual depression, might be succeeded by a layer of mud or sand, being thus sealed up, to become in after ages a bed of coal. Marine productions would not again be present in the deposit, until the depression was deep enough to bring the basin below the level of the sea ; but as the supply of detritus would be from the land, the constant accumulation of this would again shut out the sea ; and in this way alternations of strata, characterised by the presence of marine and fresh water remains, might occur precisely as we observe to be the fact in nature. Nay we ought not to be surprised, if we were to find in the same bed, alternate layers of fluviatile and marine exuviæ ; it is only to suppose, that the sea was shut out whilst a thick bed was being deposited, by which we might have marine remains at the bottom, and those of fresh water at the top, with a mixture of both between ; or the Fossil characters of the bed may be reversed, if we suppose the great depression which brought in the sea, to have taken place whilst a thick stratum was in the act of deposition.

It further appears to us that this view of the mode of formation of our Fossiliferous rocks, is borne out by the nature and condition of their organic remains—the whole of which, animal as well as vegetable, we think, prove that the beds containing them have been formed in water of a moderate depth, and as it is abundantly established that there are alternations of marine and fluviatile beds, even in deep-seated strata, we must either account for these, by a system of depression, such as we have been advocating—by which each bed was deposited so near the surface, as not to be unfavourable to the existence of animal life—or call to our aid the theory of oscillation ; as we imagine few Geologists will be found at the present day, to advocate the old notion that the organic remains in these alternating beds of such differing natures, had all been precipitated to the bottom of the same ocean. To any one who may be found entertaining such a notion as that the carboniferous group of rocks of the counties of Northumberland and Durham, for instance, were formed in a deep basin, we would observe, that it is one whole unbroken series of strata, certainly not less than 4 to 5000 feet thick. —Is it possible to suppose that the Corals, the Crinoidea, and various Testaceous Mollusca—the remains of which we find in the deepest seated Limestone and Shale beds in profusion, could exist at so profound a depth ?—or can we suppose the Vegetables,—Stigmaria, Sigillaria, &c. whose remains exist in some of the lowest Sandstone beds, alternating with these same Limestones, and which beds do not contain a vestige of marine life, could have found their way to the depths of such an ocean, without mixing with the remains of animals which existed in it in abundance? or, could portions of the most delicate Ferns, which, in point of preservation, rival the skill of the most accomplished botanical preserver, and which we find in almost the very lowest of our secondary formations, have found their way to such a situation uninjured?

The great north of England Coal-field contains 25 seams of coal, alternating with beds of Shale and Sandstone, forming altogether a thickness of 8 to 900 feet—is it possible to suppose the vegetable matter, which constituted these coal seams, to have been washed by floods from the land, and sunk to so profound a depth; every layer persistent over the whole space, and of an even thickness?—or supposing such sinking to have been possible, could the vegetable matter be free from any foreign admixture as we find the coal to be?

It is the opinion of Geologists, that most of the Saurian animals whose remains have been discovered, inhabited shallow water, in the immediate neighbourhood of land upon which they occasionally lived—as the animals nearest allied to them in nature now do—and it is known to be the general character of living testaceous Mollusca, to inhabit shoals of moderate depth, round coasts, rather than the deep sea;—this being the case, it is not unfair to suppose their ancient prototypes to have had like habits. We are aware that it is even yet supposed that the originals of all the organic remains we find have been shifted and washed about in the water,—drifted in fact far away from the spots where they lived. We have elsewhere in this work given reasons, which to our minds are conclusive, why this could not have been with vegetables;—the remains of animals of one species occurring so commonly in the carboniferous formation, where their shells are found congregated together in thousands, of all ages and sizes, we think attest the same fact:—but in the Limestone of Burdiehouse the organic remains prove incontestably, that this was the case, for we not only have them of different sizes and ages, some perfect (as many of the smaller fish, retaining every scale), but in the fæcal matter which is found abundantly disseminated throughout the whole bed, and from the examination of which we gain such a curious insight into the habits and economy of these animals, and which would be of a nature incapable of being moved without dispersion, we are incontestably led to the conclusion, that these animals lived and died in waters and near the spots where their remains now exist.

Along with Coprolites, we find in the utmost profusion, so as even in some parts to make the rock appear almost Oolitic, the shells of Cypris, with other minute Entomostraca, whose habitat is one of stagnant waters;—upon these undoubtedly the smaller fish had fed, as their existing representatives now do—they, in their turn, becoming the prey of the larger kinds.

Were any other arguments necessary to convince us that the deposition of even our deepest seated Fossiliferous beds, had taken place in comparatively shallow water, the Ripple marks so common upon them, which many Geologists think are proofs of a deposit at or near the edge, of comparatively tranquil water; and the oblique bedding of Sandstones, which is taken to indicate the quiet deposit of sand, in thin layers, over lines of surface differing from those of stratification, would furnish us with additional evidence. Both these phenomena occur in many beds, even to the very lowest, in the carboniferous formation.

The basin-shaped depressions we have been considering, would appear to offer the conditions which have, *a priori*, been thought to be necessary, to account for the tranquil deposit of most vegetable remains, but particularly those so frequently found in the carboniferous series to stand vertically across the strata;—for whether we consider these to have been drifted from their places of growth, and to have settled with their root ends down, or that the Fossils now occupy the spot on which the plants grew, as we believe many of them to do, we equally require a quiet deposit, in a situation removed from the destructive action of large masses of water in motion.

This view of gradual depression recommends itself further to our minds by its simplicity. It is difficult to conceive, in the series of strata we have had under review, that after the deposit of the Burdiehouse system, a sudden sinking of surface should take place, to receive the marine Limestone of Gilmertoun, which succeeds it in the section—this to be as suddenly brought up to be covered by the alternating Coal series of Loanhead. Difficult as this is to understand, it is however a simple problem, compared to the numerous sinkings and risings that would be necessary to account for the many and sudden alternations of salt and fresh water remains, which occur in the newer Fossiliferous rocks ; all which oscillations have taken place without leaving evidence of any peculiar disturbance in the strata supposed to indicate these movements; indeed so quiet must they have been, as not even to disturb the plane of stratification, or alter the chemical nature of the deposit, that was going on when the change took place.

161

LEPIDODENDRON LONGIFOLIUM.

Lepidodendron longifolium. *Ad. Brongn. Prodr. p.* 88.—— *Sternb. Flor. du monde primitif. t.* 3.

From the Newcastle coal measures.

Although this is not to be compared with the beautiful specimen figured by Count Sternberg, which was $2\frac{1}{2}$ feet long, with leaves 18 inches in length, yet it is interesting as shewing this remarkable plant in a new state. In Count Sternberg's specimen the branch was nearly cylindrical and very slender, but here it is more thick and compressed, as if it had been distending into a cone, or something of the sort.

In its general aspect it resembles very much *Pinus palustris* or *longifolia*, but it appears to have had its leaves growing solitary and not in pairs and clusters, and therefore could not even have belonged to the genus Pinus. In all probability it was another form of that extinct race which held a middle place between Lycopodiaceæ and Coniferous plants, as we have already explained. See *t.* 98 and 99.

LEPIDOSTROBUS COMOSUS.

From the Burdiehouse limestone.

Apparently a distinct species of this genus, differing from the other published kinds in its much larger size, in its conical figure, and in the very shaggy appearance of its outside. It is impossible to say positively what it is that has produced this appearance, but it is probable that it is owing to the great length of the points of the scales of which such cones consisted.

Drawn from a specimen belonging to the Royal Society of Edinburgh.

163

LEPIDOSTROBUS ORNATUS; var. *didymus*.

See tab. 26.

The beautiful specimens from which this drawing has been taken were communicated to us through Professor Graham by Lord Greenock, who found them in the Ironstone at Newhaven, near Edinburgh.

They are an instance of apparent malformation, in consequence of two cones having grown together, but they do not throw any more light upon the real structure of Lepidostrobus than has already been given at t. 26 of this work. They are, however, so extremely beautiful, in consequence of the skilful manner in which they have been polished, that we gladly seize an opportunity of making them known.

It would seem as if a Palm leaf had been pressed over the outside of the cone at fig. 1.

PINUS ANTHRACINA.

From a single specimen found in the Coal measures of Newcastle by Mr. Buddle this figure has been taken. It is too imperfect to enable us to form any other opinion concerning it, than that it was the cone of a Fir, which in all probability, belonged to the modern genus Pinus, if we are to judge from the great thickness of its scales.

We know no modern species with which it would be of any use to compare it.

ZAMIA GIGAS.

One of the largest of this genus found in the Oolitic rocks of Scarborough. It has occurred as much as three feet in length. Mr. Williamson, Jun., whose drawing we use, informs us that its leaflets are sharp pointed, with regular veins which are simple, and, like those of most monocotyledons, terminate at the narrow apex, though some of them have formed little points on the margin of the leaflets. The latter circumstance may be considered to indicate the commencement of lateral teeth, and if so the identity of this and similar remains with modern Zamias will be more strongly than ever demonstrated; because Zamia is the only modern Cycadeous genus known in which the leaflets have lateral points.

If it were not for the manner in which the veins are distributed, we should have supposed this to be the same as the Cycadites latifolius, figured by Professor Philips at t. x. f. 3, of his second edition of the Geology of the Yorkshire Coast. But in that plant the veins are represented as losing themselves in the margin along the whole of the upper edge of the leaflets, and not terminating in almost all cases in the apex.

STIGMARIA FICOIDES.

(Its Anatomy.)

See tab. 31 to 56.

We have so frequently referred to this extraordinary fossil already, and have so continually insisted upon the impossibility of forming any opinion of what it really was, that we have peculiar satisfaction in being able at last to prove the correctness of our supposition, that it was in reality a plant of which no modern analogue exists, by its anatomy. For this we are indebted to our friend Mr. Prestwich, who placed in our hands some time since a fragment of Stigmaria, preserved in Ironstone from Colebrooke dale, which seemed to promise some preservation of tissue. Upon being polished by Mr. Cuttell, an ingenious London lapidary, and a worthy rival of Mr. Sanderson of Edinburgh, it presented the appearances here represented.

The transverse section exhibited a meshing something like that of Coniferæ, but with no concentric circles, and with the medullary rays consisting rather of open spaces between the other tissue, than of the common muriform tissue found in such places. The longitudinal section (*fig.* 2.) presented an assemblage of spiral vessels, of a very tortuous and unequal figure, without any woody or cellular matter intermixed.

These formed a cylinder which was surrounded externally by a mass of inorganic mineral matter, upon whose surface the peculiar markings of Stigmaria were preserved, and which enclosed a hollow cavity altogether destitute of mineral deposit.

It would therefore appear that Stigmaria was a plant with a very thick cellular coating or bark, surrounding a hollow cylinder, composed exclusively of spiral vessels, and containing a rather thick pith; and that the plates of cellular tissue which preserved the communication between the bark and the pith were of so delicate an organization, that they disappeared under the mineralizing process which fixed the organic characters of the wood.

It is almost needless to say that no plants of the present day have such a structure, and that consequently our original impression that Stigmaria represents an altogether extinct race was correct.

We must, however, remark that we strongly suspect our *Caulopteris gracilis*, (t. 141.) to be the same thing, as well as Mr. Witham's *Anabathra pulcherrima*.

THUITES EXPANSUS.

Thuites expansus. *Sternb. Fl. der vorw. t.* 38. *fol.* 1, *&* 2. *Phillips Geol. Yorks. t.* 10, *f.* 11.

From the lower Sandstone shale and coal of the Oolitic beds near Scarborough.

Mr. Williamson, Jun., remarks that this plant, " in its general form and mode of branching, bears a considerable resemblance to Lycopodites Williamsonis, but differs from that plant in the scales being shorter, broader, and more flattened, and also in the absence of stipules. It is seldom found large:

I never saw it more than twice the length of the drawing. The scales have apparently been equally attached to the stem on every side. They are nearly as broad as long, and furnished at the apex with an obtuse mucro. There is a slight elevation at the centre, which nearly disappears at the base, and there is an indistinct marginal depression. The species resembles in a remarkable degree Brongniart's *Fucoides Brardii.*"

This is undoubtedly true; and we cannot but wonder that so good a Botanist as Adolphe Brongniart should have referred such remains to the tribe of sea weeds. The species is so extremely close to Coniferous plants related to Callitris and Dacrydium that, although we are unable to identify it with any existing plant, yet we cannot suppose that it was not nearly allied to them, and we do not see how it is to be distinguished from the *Fucoides Brardii* already mentioned, which is from beds of lignite below the chalk.

It scarcely answers to the genus Thuites, its branches having apparently been round instead of flat; but we leave it with the name it already bears, in the hope that its true place will be determined by Adolphe Brongniart in his general view of Fossil Coniferæ.

The intramarginal veins and very prominent midrib of the leaves, are probably owing to shrinking after long maceration in water.

SPHENOPTERIS ARGUTA.

From the Oolitic rocks of Scarborough. Supplied by Mr. Williamson, Jun., with the following note.

" A bipinnated frond of small size, but of remarkable elegance. The central rib is prominent but bears no marks of any angles. The leaflets are rhomboidal, the lower ones with from 5 to 9 lobes, which are generally split at their ends, the lobe at the base of each leaflet being generally the largest. The veins are very indistinct, but certainly forked, one branch entering each lobe. Towards the apex of the frond the leaflets are very small, but once or twice lobed, and that on the upper edge only. In the lower leaflets the lobes are almost always opposite, and the connecting portion between each pair but little more than the diameter of the central rib. We have no species of Sphenopteris with which this can be compared. It is decidedly different from *Sph. stipata,* as well as *Sph. hymenophylloides,* the latter of which in particular has shallower lobes to the frond."

174

NEUROPTERIS ATTENUATA.

A coal measure fern, allied to *Neuropteris Loshii*, from which and all other species it differs in its leaflets becoming gradually smaller, till the terminal one is less than any of the others. This is an unusual circumstance in the genus Neuropteris and distinctly marks the species.

175

ZAMIA TAXINA.

From the Stonesfield slate. Communicated by Dr. Buckland.

It occurs in specimens no larger than those now figured, and is so nearly allied to Zamia pectinata as to look like a small state of it. We are by no means assured that it really is not so, but its leaflets seem to be a little less approximated, and more gradually tapered to a point; and these circumstances, together with its size, induce us to look upon it as a distinct species.

176 A

SPHENOPTERIS CYSTEOIDES.

From the Stonesfield slate; communicated by Dr. Buckland.

A single ferruginous impression preserved in Sandstone is all that we know of this plant, which seems to have been bipinnated with ovate lanceolate acuminate divisions and pinnatifid segments. It is very like a bit of the recent Cystopteris fragilis, but it is quite impossible to form any opinion as to whether or not it was distinct from that species.

176 B

TÆNIOPTERIS VITTATA.

Tæniopteris vittata. *Supra, vol.* 1. *t.* 62.

From the Stonesfield slate; communicated by Dr. Buckland.

This is apparently the very specimen figured by Sternberg, and upon which the species is founded. Is it really the same as the plant from the shale of the Gristhorpe bed, and already figured vol. 1. tab. 62. of this work? We suspect not; the leaf is broader, the leaves more closely veined, and the aspect of the impression that of a plant with a more leathery texture. We have, however, no positive means of judging, and it is very possible that the Gristhorpe specimens are young, while that now figured is old and matured. A further examination of the Stonesfield slate will alone decide the point.

SPHENOPTERIS HIBBERTI.

From a specimen, for which we are obliged to Dr. Hibbert, who procured it from an interesting deposit of fresh-water Limestone occurring at Kirkton near Bathgate, in the county of Linlithgo.

This is described in a memoir upon certain fresh-water Limestones, published in the Transactions of the Royal Society of Edinburgh, vol. xiii. from which we extract the following.

" A mile or two to the east of Bathgate, at Kirkton, we find that a very considerable outbreak of greenstone has occurred. Close to it on the west appears the limestone of Kirkton. By this contiguity, we are assured, that the limestone must have been elaborated within the immediate sphere and influence of an extensive volcanic eruption. The consequence has been, that one of the most unique formations of which Great Britain can boast, has been formed, indicative of thermal waters, belonging to the carboniferous epoch.

" A decidedly fresh-water formation is thus exposed, which is characterised by the absence of all marine shells, corallines, &c., and the presence of the well known vegetable remains of the Coal formation.

" But the remarkable circumstance in this limestone is its mineralogical character, indicative of the very powerful chemical action under which it was elaborated. This chemical action appears to have been so energetic, as to have caused such miscellaneous earthy matters as are found to enter into the composition of an impure limestone, like that of Kirkton, to separate into laminæ, and to assume a sort of striped disposition (*rubané* as it is also named) resembling what I have occasionally noticed in Auvergne, where tertiary strata have come into contact with volcanic rocks. The strata, for instance, of Kirkton quarry, are composed of distinct and alternating thin laminæ, some of them being of remarkable tenuity, variously consisting either of pure calcareous matter, of translucent silex, resembling common flint, or of a mixed, argillaceous substance, which approaches to the character of porcellanite, or of ferruginous, or even of bituminous layers, originating probably from vegetable matter.

" Upon one of these very thin aluminous folia, which I have compared to porcellanite, I observed the impression of a Fern, apparently of a Pecopteris, which was delineated upon it like a painting upon porcelain "

In many respects the specimen resembles Sphenopteris polyphylla of the Clee Hills (fol. 147), but it appears to differ in the following particulars. Its leaflets are less regularly three-lobed; when they are so lobed the central segment is scarcely different in size from those at the sides, and the latter taper more gradually into the stalk. These circumstances produce a considerable difference in the general appearance of the two plants, although they may not seem upon paper to be of much weight.

178

SPHENOPTERIS LATIFOLIA.

Sphenopteris latifolia. *Supra, vol. 2. p.* 156.

From the Bensham coal seam in Jarrow Colliery.

It is most difficult to form a correct opinion of what are distinct species and what are merely different parts of the leaf of the Ferns found in a fossil state, so much do the different portions of the leaves of recent species vary between their base and their apex; a property which we shall presently see was at least as strongly characteristic of the species of the ancient Flora.

At fol. 156 is represented a plant called Sphe nopteris latifolia, which appears to differ from that before us in the ultimate leaflets being scarcely ever three-lobed, but usually consisting of from five to seven divisions; here they are almost constantly three-lobed, the instances of five lobes being extremely rare. We are however persuaded that in reality the present plant with its 3-lobed leaflets is a portion of the upper end of a 5 or 7-lobed state, and it is far from improbable that Sphenopteris dilatata (vol. i. tab. 47) is again the extreme point of the leaf of such a species with leaflets altogether undivided or merely two-lobed.

At all events no safe geological inferences can be drawn from the presence of such remains; that is to say, supposing the plant figured at tab. 156 and the present, and that of tab. 47 were all found in separate stations, no one ought to consider such a fact of the slightest weight in shewing the three stations to be different formations.

PECOPTERIS LOBIFOLIA.

Neuropteris lobifolia. *Phillips Geol. Yorks. t.* 8. *f.* 13.
Neuropteris undulata. *Supra, vol.* 2. *t.* 83.

Three specimens only of this plant have occurred in the lower sandstone and shale of the Oolitic formation at Haiburn Wyke near Scarborough.

Mr. Williamson, Jun. from whom we have received the drawing has made us the following communication upon the subject.

" This bears a strong resemblance to *Neuropteris undulata,* figured by you some time ago; belonging to which species I have also discovered a new character I had not observed in the specimen drawn, owing to its imperfect state. It consists in a large lobed leaflet at the under part of the base of each pinnule, and is consequently the same species as that of which Professor Phillips has given a drawing under the name of *Neuropteris lobifolia.*

"The present species differs from that plant in having these lobed leaflets as well as the others much smaller, the lobes, although variable, more of an hexagonal form, and the nervures much fewer in number.

"Fig. 2 is a magnified view of the base of a pinnule of the present plant, and fig. 3 represents three of the leaflets from the extremity of a pinnule, the apex of each being broken off. Fig. 4 shows a similar portion of the original Neuropteris undulata."

Upon considering these additional points of information, and the drawing now published, it is in the first place apparent that the species is a Pecopteris and not a Neuropteris. For there is a manifest midrib, and the lateral veins are planted upon it abruptly; and secondly, that it is closely related to the P. acutifolia and obtusifolia represented at tab. 157 and 158 of the present work.

With regard to the differences pointed out by Mr. Williamson, we conceive they are not greater than might be expected upon different parts of the same leaf; and that while tab. 83 and 179, fig. 4, represent portions of the lower part of a leaf, t. 179, fig. 1. 2. 3. belong to the upper part of it.

180

ASTEROPHYLLITES TUBERCULATA.

Supra, vol. 1. *t.* 14.

From the roof of the Bensham coal seam in Jarrow Colliery.

We figure this fragment with a view to completing the representation given in the place above referred to, and to shewing that it is in all probability the remains of a mass of inflorescence, of which we have here a portion of the naked stalk. But we are still uncertain as to what it may have been or belonged to. There is some room for suspecting that it was a part of the fructification of Calamites, see tab. 15-16; but after a lapse of nearly five years we are as much in the dark upon this subject as ever.

It is wonderful that no one should ever yet have been able to find Calamites in actual connection with such remains as this; and that being the fact the probability of the two belonging to each other is by no means increased.

181

SPHENOPTERIS FURCATA.

Sphenopteris furcata. *Ad. Brongniart, hist. des Vég. foss. t.* 49. *f.* 4. 5.

From the roof of the Bensham coal seam in Jarrow Colliery.

Another of the large family of narrow-leaved Sphenopterides, to which S. affinis, artemisiæfolia, crithmifolia, &c. belong. It seems wholly impracticable to define the species, if species they be, with any precision; and it is probable that many of them will be one day considered mere forms of each other.

Newcastle, Charleroi, and Saarbruck are given by Mons. Adolphe Brongniart as localities for this species, which differs from S. crithmifolia, chiefly in its more compact mode of growth, and in the general outline formed by the points of the lobes of the pinnules being more broadly oblong.

PINUS CANARIENSIS.

We trust we shall stand excused for publishing this remarkable cone in the Fossil Flora, although it does not actually belong to the deposits of this country. We are the more induced to venture upon this departure from our original plan in consequence of the numerous other cones which have been already figured in this work.

It was found in Spain in the year 1832, by Colonel Silvertop, in a deposit of indurated whitish marle, containing powerful beds of sulphur, near the town of Hellin, in the province of Murcia. Impressions of fish are found in the same locality, which Professor Buckland considers a tertiary formation.

What this cone may have been in an age so nearly approaching to the present order of things becomes a highly interesting point.

The specimen from which our drawing was taken

is of a pale dirty white, (not black as the heavy shadows in the plate would seem to indicate), six inches long by three wide in the broadest part; the lower part, however, having been broken away, it must have been originally something longer. It is so much compressed as not to be more than two inches thick in the flattened direction. The scales of which it was composed were about two inches and a half long, and terminated in a broad, elevated, woody point, which, although very much broken, appears to have curved backward; the sides of the lozenges formed by the junction of the ends of the scales, were almost thirteen-twentieths of an inch long. A longitudinal section of the cone does not enable us to judge of what nature the seeds were.

Certainly there is now no European Pine with which this can be identified. The only two that approach it in size are those of Pinus Pinea and P. Pinaster. The former has a much rounder figure, and the ends of the scales are not recurved. No one ever saw P. Pinaster with cones more than half the size of this, and their figure is conical and a little curved, not oblong.

If we consider extra-European species, we shall find it approach in several respects, especially in size and general form, the Pinus longifolia of India, which has the points of the scales of its cone recurved; but the scales of that species do not form regular lozenge-shaped spaces by their junction; on the contrary the two upper sides of the lozenges are very much shorter than the two lower.

But it is with a species which now forms a tree of considerable size on the mountains of Teneriffe and the Island of Grand Canary, that we would especially compare the fossil. That plant, the P. Canariensis of Mr. Lambert's Monograph of the genus Pinus, vol. 1. t. 28, (8vo. edition), has cones which seem in reality to be identical with this, in size, general figure, and the form of the ends of the scales. Whether or not the nature of the seeds was also the same we have no means of ascertaining; but as far as the evidence goes we are unable to discover any Botanical distinction.

This is an interesting circumstance, as affording an instance of the remains of an existing species of plant in a tertiary deposit.

NEUROPTERIS HETEROPHYLLA.

N. heterophylla. *Ad. Brongn. hist. des Vég. foss. t.* 71. 72. *f.* 2.
N. acutifolia. *Ad. Brongn. hist. des. Vég. foss. t.* 64. *f.* 6. 7.
Filicites heterophylla. *Ad. Brongn. class. vég. foss. t.* 2. *f.* 6.
Pecopteris De Thiersii. *Id. Prodr. p.* 56.

For this remarkable specimen we are indebted to Mr. W. D. Saull, who purchased it out of the Museum of the late Mr. Sowerby. It lies in a large nodule of Ironstone, and is in a singularly perfect state. That it belongs to some of the old Coal-measure formations admits of no doubt.

It was probably the end of a leaf of some Fern of large dimensions, and seems, from its convexity, to have been originally of a tough and thick consistence. What is most remarkable in it is that the leaflets of the very same portion of a leaf should be so extremely different on opposite sides of the same rachis. On the right, the pinnæ are about two inches long, cordate, a little wavy, and altogether undivided; while on the left they are three inches long, deeply pinnatifid into about seven pairs of ovate lobes, and a terminal one resembling the entire pinnæ on the opposite side, except in being much smaller.

Hence it has been well named *N. heterophylla*, by Mons. Adolphe Brongniart, who figures it from the coalfields of Charleroi and Saarbruck. It appears to us that the *N acutifolia* of the same Botanist, from Wilksbarre in Pennsylvania and from Bath, also consists of the entire pinnules of this same plant; and that it is even doubtful whether *N. Loshii*, *angustifolia*, and *Scheuchzeri* are not all fragments of this same species.

PECOPTERIS ABBREVIATA.

P. abbreviata. *Ad. Brongn. Prodr. p.* 58. *Vég. Foss.* 1. 337. *t.* 115. *fig.* 1—4.

From the coal measures of Welbatch near Shrewsbury, communicated by Mr. Corbett. It is described by M. Adolphe Brongniart, from the Bath coal field and from the mines of Anzin near Valenciennes.

Among all the species of Pecopteris this is known by the pinnæ being merely crenelled, with from three to five veins occupying the middle of each crenature. It appears from M. Adolphe Brongniart's figure that the number of the veins is sometimes increased, but we have not remarked that circumstance in the specimen before us.

FUCOIDES ARCUATUS.

From the collection of Professor Phillips, who states that it is the only one he has seen from the extinct flora of Gristhorpe.

It does not appear referable to any of the species figured in Brongniart's Végétaux fossiles, neither does it approach any modern species, so far as it is possible to judge from the fragment before us. We do not however see any reason to doubt its being a Fucoid, and that is all that can be said about it.

We presume the white spaces which divide each lobe of the plant into two parallel portions, represent the place where a thickened midrib once existed.

EQUISETUM LATERALE.

Equisetum laterale. *Phillips's Geol. of Yorks. t.* 10. *f.* 13.

Found not uncommonly at Haiburn Wyke in the lower sandstone and shale of the Oolite; also at White Nab, on the coast of Yorkshire, south of Scarborough, whence the specimen now represented was communicated by Professor Phillips.

It always occurs in small pieces, rarely exceeding in size that of our drawing; a fragment with 3 or 4 articulations is considered unusually perfect. The stem itself is destitute of furrows; but at the articulations it seems as if a striated sheath, or else a whorl of short fine leaves could be traced. And what is most remarkable, at irregular distances between the articulations are found little round disks with lines radiating from a common centre, something in the way of the phragma of a Calamite. These disks which look like the scars left behind by branches that had fallen off, are not stationed at the axils or the articulations, but appear at uncertain points on the internodes, and according to Mr. Williamson, Jun. are found less frequently on the stem than loose in the shale, without any apparent connection with the plant. This is a singular fact, and would lead one to think that the disks hardly belong to the stems with which they are found associated.

We retain Mr. Phillips's name, in consequence of the great obscurity that attends the species; but we may remark that we are by no means satisfied that it is the remains of an Equisetum.

PECOPTERIS HAIBURNENSIS.

Communicated by Professor Phillips from Haiburn Wyke on the east coast of Yorkshire, where a small colliery is worked in the lowest beds of the Oolitic series.

It appears essentially distinct from all the species hitherto figured of this extensive and difficult genus, but it approaches the P. pteroides of the coal measures. It is apparently distinguished however not only by its greater size, but also by its more membranous texture, and by the lowest pinnule of each pinna not being placed very obliquely in the angle formed by the separation of the partial from the common rachis.

In the magnified representation in the accompanying plate, the artist has drawn the separate pinnule as if it adhered to its rachis by the centre only; but we believe this appearance to be caused by a decay of a portion of the base of the pinnule.

In general the pinnules adhere by their whole base, so that the pinnæ are pinnatifid rather than pinnate. This is correctly shown in the principal figure.

188

BRACHYPHYLLUM MAMMILLARE.

Brachyphyllum mamillare. *A. Brongn. Prodr. p.* 109.

Communicated by Professor Phillips, from the Oolitic coal field of Haiburn Wyke, in Yorkshire.

M. Adolphe Brongniart considers this as a doubtful Coniferous plant; but it must be confessed that it has as strong a claim to be received as an unquestionable species of the Pine tribe, as any other fossil of which the leaves alone are known.

Its general appearance, and the arrangement of its leaves, which are all that we have to guide us in forming an opinion of its analogies, are quite those of the Coniferous plants allied to Araucaria excelsa, Callitris and Dacrydium.

Fig. 1. 2. 4.

CARPOLITHES CONICA.

From the Coralline Oolite of Malton, whence Mr. Williamson, Jun. has sent specimens to Professor Buckland, and drawings to ourselves. We have also seen it from Dr. Murray of Scarborough.

It is the remains of some fruit. The specimens are conical and three-sided, with the base(?) convex and bordered with a single row of tubercles, and divided by three elevated ridges, while the sides are perfectly smooth. Sometimes the three-sided character is absent, and the specimens are merely flattened, with an elevated edge on each side, and an elevated line passing through the truncated end. The greater part seem to have lost their external rind, but in one specimen, communicated by Professor Buckland, the sides are crushed and exhibit a cellular appearance here and there, as if portions only of the rind had rotted away.

The form of this fossil obviously calls to mind the genus Trigonocarpon, which we have already shewn to be undoubtedly a Palm fruit. But as we are unable to discover any trace of the necessary evidence as to this in the specimens under consideration, we think it best to refer them to the provisional genus Carpolithes.

Fig. 3 and 5.

CARPOLITHES BUCKLANDII.

Carpolithes Bucklandii. *Williamson fil. Mss.*

Communicated with the last by Mr. Williamson, Jun. from the Coralline Oolite of Malton.

It appears to be of the same nature as the subject of the last article, but to differ specifically in its much more oblong figure, in the smallness of the truncated end, and in the sides being distinctly tuberculated as well as the edges of the truncated end.

Mr. Williamson describes the specimens thus:

"They are generally of a rounded or rather flattened form. The base, or part of attachment, is surrounded by a ring of prominent but often irregular tubercles. The space within the circle is generally divided in three by elevated ridges, which I can compare to nothing so aptly as to the lines of carbonate of Lime in a nodule of Septarium. But these lines rarely pass beyond the circle, and in No. 3. do not exist; a slightly elevated tubercle forms the centre. The other parts of the fossil are covered with very irregularly placed tubercles, the spaces between which are quite smooth. In none of the specimens I have seen, which have not been few, was any appearance of a division into lobes. The specimens are of a deep brown colour."

190—191

HIPPURITES LONGIFOLIA.

At t. 114 of the second volume of this work is figured the original species of this genus, from the Newcastle coal-field.

By the kindness of Mr. Arthur Montague, of Park End near Lydney, in the Forest of Dean, we are furnished with the accompanying representations of what seems another species, specimens of which exist in the collection of himself and Mr. Henry James. They occurred in the shale of the Forest of Dean coal basin.

This appears to differ from H. gigantea in the leaves being longer than the spaces between them, and the stem being quite smooth.

Fig. 190 represents some fragments of the stem and leaves of the natural size; fig. 191 is a diminished figure of a considerable portion.

FAVULARIA NODOSA.

Favularia nodosa. *Bowman Mss.*

For the drawing and following account of these remarkable remains, we are indebted to J. E. Bowman, Esq. of Gresford near Wrexham.

"From the roof of the lowest bed of Coal at Flint Marsh Colliery, on the estuary of the Dee, among abundance of Sigillariæ and Calamites of large dimensions.

"This beautiful fossil is in fine soft shale, and retains on one side the carbonized exterior surface of its vegetable form. The undulations and pencillings of the areolæ, to which the basis of the leaves have been attached, are as clear and sharp as the impression from a seal, and even require the lens to shew their delicate inequalities. These areolæ cover the whole surface, are wider than their length, and not only touch the intermediate furrow, but give it a waved character, by encroaching upon the contiguous row beyond it, right and left. They are separated from those above and below in the same row, by slightly and uniformly diagonal lines or septa, which throw them into rhombs. In this specimen the surface is flat, but in others, detached portions of the rows have a considerable elevation, with the intermediate furrow deep in proportion. Where this is the case, the shale is seen to be more strongly impregnated with iron, which circumstance may have caused it *to set*, or harden quicker from its soft state, and so, more effectually to resist the superincumbent pressure. In the elevated parts, the transverse divisions of the rows, are nearly at right angles with the axis of the stem, which shews that their diagonal or slanting direction in the flatter parts, is the result of pressure, and not of a spiral arrangement of the leaves, as in Lepidodendra and modern Coniferæ. This will be evident by observing that the diagonal septa *ascend* from right to left on one side of the fossil, and from left to right on the opposite one; while all spirals ascend in the *same* direction, on which side soever they are viewed, as the stem of the Hop or Kidney-bean on its pole, or a thread wound round the finger from the base to the tip; and these fossils are only flattened cylinders.

"In its decorticated portions, the present somewhat resembles Favularia tessellata, already figured at plates 73 and 74 of this work ; but the furrows, from the cause above mentioned, are somewhat wavy or zigzag, the areolæ rhomboidal and shorter in proportion to their width, and the central scar, or vascular connection between the leaves and the interior of the stem, of an elevated circular or horseshoe form, around and within which the general surface is depressed. In F. tessellata this portion is shaped like the club upon playing cards, with the central lobe elongated, and reaching nearly to the apex. It was not clearly stated in describing that fossil, that the engravings shew its *decorticated* appearance, and that the exterior carbonized surface is in no part well preserved.

"The irregularly warted zone which crosses the central portion of the fossil, interrupting the longitudinal rows, gives the idea of a joint, or division of the stem. As this mark occurs in a precisely similar situation, on both sides of eight or nine other specimens, and as no instance of a jointed structure has yet, (as far as I am aware) been found in any Favularia or Sigillaria, it may be worth while to examine how far it goes to support that character. I once thought it might be the effect of a twist or bruise the vegetable had received while in a soft state ; but a more attentive examination, shewed that some such accident had actually happened in another portion of the decorticated surface, and had left the impression of an irregular transverse rent, with a sharp jagged edge, but without affecting the opposite side. I am therefore satisfied from this, and from the smooth unbroken undulations of the zone, that it is a part of the vegetable itself. Its inequalities have, however, been more strongly impressed on its interior than on its exterior surface. Indeed the areolæ, though much transposed, and less densely imbricated in these parts, are never absent, and where they retain their carbonized surface, there is no exterior trace of joint or sheath, as in Calamites or Grasses. Again, in all the specimens, the longitudinal rows of areolæ, after having been more or less thrown out of their perpendicular direction, in the neighbourhood of the zones, soon regain their natural position. In two or three instances, however, an additional row is inserted, the space for which has been made by the bend or divergence of several of the collateral rows on each side, from their parallelism; but neither the one nor the other is narrower than the rest, in consequence of this intrusion. Lastly, in all the specimens before me, the zones or joints occur on both surfaces, and intersect the axis of the stem, at precisely the same angle as the transverse divisions of the areolæ. Though I have shewn that this diagonal direction is the result of pressure, the coincidence of the joints is a fact of great importance as pointing to a common cause in the structure of the plant, and proving their original position to have been at right angles with the axis of the stem. On the whole it appears highly probable that these zones indicate the jointed structure of such plants as Favularia (and perhaps we may add, from general analogy, of Sigillaria also) but that the joints were not accompanied by any other external appearance than a less densely imbricated and perhaps a longer foliage, and an interruption of the parallelism of the rows. As there is no trace of ramification in any of the specimens, it occurred to me whether these joints might be the rudiments of a whorl of branches, as in recent Coniferæ; but the bases are not broad enough to render this probable. In one instance only, two joints occur in the same specimen, viz., one at each end, separated by an interval of 4½ inches, which may help to convey an idea of the real character and habit of these wonderful vegetables. If we suppose for a moment that the joints were ramuli, or whorls of longer foliage, the plant would not be very dissimilar to a gigantic Equisetum or Hippuris, whose stem was concealed by leaves or scales.

"The specimens are all completely covered on both sides with areolæ, arranged in perpendicular rows, which seem to lock into each other, and to form a rich mosaic work. It is singular that one surface should retain its carbonized exterior, and the other be for the most part decorticated. Yet they all are so. They are also generally broken off at the joints, but there is no internal trace of a phragma, or of a central wooded axis. As the specimen figured (which is 10 inches long by nearly 6 wide) is not half an inch thick, and not more than a quarter of an inch about the edges or sides, the absence of an axis goes far to prove that such plants were destitute of that arrangement. And if they had a jointed structure, the nodi must have been soft and perishable, as the thickness of the fossil is not greater where they occur.

"I am not aware that either the leaves or fruit of these primæval vegetables have ever been found, or at least identified. From the evidence we have of the abundance of the former, we are almost driven to the conclusion that they must have been supposed to belong to some other fossil, or that they have been of a soft and succulent nature, and have yielded to the wasting causes which the tougher material of the stem has resisted. Is it an improbable conjecture, judging from the very broad thick base by which they were attached to the stem, that in mature plants, they each formed an involucre, enveloping one or more seeds, like the leaves of Isoetes, or the imbricated scales of the cones of the Pinus family?"

---

a. Shews the exterior surface, which is of coal, as beautifully sharp and perfect as the finest impression of a seal. The inequalities of the surface are but slight; but some of the more delicate lights and shades are so fine, I have not attempted to express them.

b. A portion of another specimen of the same fossil, shewing the interior surface or decorticated state, after the removal of the coaly shell—In this specimen the leaves are not quite so closely set.

c. Portion of the same specimen as *a*, remarkable for the different arrangement of the leaves in the central part, where they seem to be disposed spirally or alternately, and in the specimen itself give the idea of a joint; this sketch shows a part of the exterior leaf-bases, and some of the decorticated portion.

A. 1, 2, 3, 4.

CARPOLITHES ———.

Fossil seeds from the calcareous Slate of Stonesfield, one of the subdivisions of the lower portion of the Oolitic series, so remarkable for the singular variety of its organic remains. The specimens are from the collection of Professor Phillips, who observes—" In general they may be affirmed to have much analogy with the Monocotyledonous fruit from the Oolite of Malton (Carpolithes conica, t. 189). The nourishing vessels have left prominent marks; on one flat plate is a very singular arrangement."

We find nothing sufficiently remarkable in the specimens to enable us to give even a guess at their affinity; and it is even probable that three different species are here combined, namely A. 1 & 3; A. 2; and A. 4.

193

B. 1, 2, 3, 4.

TRIGONOCARPUM NÖGGERATHI.

This has already been published at t. 142. of the 2nd volume of this work. The specimens before us occurred in the Newcastle coal field, where, however, the fossil is rare; it is always in groups when it is met with. The specimens are less water worn than those before published, and will serve to give a more distinct idea of the species. They have been communicated to us from Holywell Colliery, by John Buddle, Esq., and from Wilmington Colliery, by Geo. Johnson, Esq.

193

C.

TRIGONOCARPUM OBLONGUM.

Communicated by Professor Phillips from the coal formation at Hound Hill, near Pontefract. It occurs in sandstone, and seems to have been buried in the midst of a mass of the leaves of Poacites cocoina.

The specimen is preserved nearly flat, and shows only two of the elevated ridges belonging to the genus Trigonocarpum. But it is obvious, upon an attentive examination, that it had three, and at the base there is a distinct 3-cornered depression.

194

ZAMIA LANCEOLATA.

From the Oolite at Haiburn Wyke, near Scarborough, whence a drawing and description have been sent us by Mr. Williamson, Jun.

"The rachis was straight, and apparently smooth. The pinnæ were long and lanceolate, contracting at their base into the appearance of a short stalk; sometimes they were opposite, sometimes alternate; their surface is covered with a series of minute longitudinal striæ, apparently simple, and not running out at the side of the leaf as in Otopteris; but the seam in which the Haiburn plants are found, being of a more micaceous nature than that at Gristhorpe, veins and delicate markings are rarely preserved."

This leaf has no doubt been produced by some one of the Cycadeoideous stems of the Oolitic Rocks, but there seems no present probability of our ascertaining by which.

VOLTZIA PHILLIPSII.

Communicated by Professor Phillips, who considers it a Voltzia, and, although a very imperfect specimen, worth figuring, because of the extreme rarity of vegetable remains in the Magnesian limestone.

In the county of Durham, where this formation is extensively developed, the remains of vegetables occasionally occur in the lower slaty beds, associated with fossil fish, and at Whitley, near the sea coast, in the county of Northumberland, we lately ascertained that a similar association occurred. Here the Magnesian limestone, which forms the whole coast of Durham, after crossing the Tyne finally disappears; an excavation made down to the lower red sandstone at this spot, exhibited thin beds of a blue calcareous shale, alternating with others of compact blue limestone. In the shale beds, the remains of fish occurred in great abundance, occasionally perfect, but generally in mere fragments; so abundant were they, that one layer of shale, not more than eighteen inches thick, in a space of ten feet square, furnished indications of several hundred individuals. In this shale-bed fragments of vegetable remains were found, but too indistinct to allow of identification; they had evidently possessed considerable substance, but were changed into a fine shining coal.

The leaves of this plant appear to have been about half an inch long, rounded at the extremity, decurrent at the back, and arranged regularly over the surface of the stem. Whether the crushed end was a portion of the fructification, the specimen does not shew; but we agree with Professor Phillips in considering it analogous to the Voltzias of the new red sandstone. It may, however, have been an Araucaria, allied to Araucaria excelsa.

CALAMITES INEQUALIS.

From the collection of Professor Phillips, who says, " This Calamites is, I think, a new species. Its irregular swellings, and unsymmetrical ramifications are remarkable. I found it with a thousand others, chiefly of the ordinary forms of Calamites, in a sandstone quarry east of Sheffield in 1827; in which quarry you might easily obtain specimens four feet long, and not much tapering; no roots are there seen, but the stems lie in all directions in the sandstone."

It is most nearly related to C. Voltzii, with which it corresponds in its lateral scars, and irregular manner of growth; but it seems to differ in its joints not regularly diminishing in size. The nodes are remarkably tumid at places, as if knots were formed in the substance of the plant; and, what is very interesting, the specimens confirm the opinion that Calamites were hollow. The cylinder that once was of vegetable matter has altogether a different texture from the interior, which is a coarse grit that separates freely from the stem itself.

NEUROPTERIS HETEROPHYLLA.

N. heterophylla. *Ad. Brongn. Prodr. p.* 53.—*Veg. foss. p.* 243. *tt.* 71 & 72. *fig.* 2.

This specimen occurred in shale, worked for the Iron-stone it contains, a little north of Whitley, on the coast of Northumberland. In the nodules of Iron-stone here, where vegetable remains are scarce, we lately discovered coprolites, containing occasionally the scales of fish; beautifully defined scales of the Heloptychius Hibberti of Agassiz, of a comparatively small size, and one fragment of a large dorsal ray, similar, but perhaps not identical, with those of the Gyracanthus found in the Newcastle coal-field, at Burdie House, and elsewhere.

The species derives its name from the diversity of form in the leaflets of different parts of the same leaf. In the specimen now figured they are all of the same oblong figure, and altogether undivided; but in other parts they are furnished with auricles at the base (*see tab.* 200.), and so acquire, when not observed in connection with each other, the appearance of entirely different species. This peculiarity is by no means confined to Fossil Ferns, but also occurs in several modern species of Pteris.

LEPIDOSTROBUS PINASTER.

From the coal at South Shields: communicated by Thomas Stephens, Esq., North Shields.

This new species of Lepidostrobus has several well marked characters that distinguish it from those already known. Its form is more narrow; its scales have a more prominent scar; their leafy points *seem* to have been more rigid; and the number of scales that the impression exhibits across it does not exceed three. The scales are rhomboidal, with a transverse oval scar at the end which is most rounded; and that end is thicker and bevelled off at the edge. In the middle of that margin of the oval scar which is next the bevelled edge is a small tubercle. Nothing is distinctly seen of the leafy points; but there are streaks and stains as shewn in the figure, which may possibly be their remains.

Although the cone is so placed in our figure as to

have its broadest end downwards, yet we are by no means certain that it is not inverted; on the contrary, as the bevelled end of the scales is at the base, when the cone is in this position, and as in recent cones it is invariably at the apex, it is probable that the greater width of one end than the other, which by the way is exaggerated in the figure, is owing to the scales of the upper end of the cone having opened a little, while those of the base remain closely pressed together.

199

LEPIDODENDRON ELEGANS.

L. elegans. *Suprà vol. 2. tab.* 118.

This beautiful impression of the surface of a large arm of Lepidodendron elegans, is from Felling Colliery. It is of the natural size, and completes the figure already given of the species in the place above referred to.

200

NEUROPTERIS HETEROPHYLLA.

N. heterophylla. *Suprà Tab.* 197.

From Jarrow Colliery.

It is here that the separation of the leaflets of this fern becomes apparent. A collector would hardly suspect the specimen now before us, and that figured at tab. 197, to belong to the very same species, and yet the beautiful specimen figured by M. Adolphe Brongniart, at tab. 71 of his work on Fossil Plants, demonstrates the fact.

The species is near *N. Loshii*, tab. 49, from which it differs according to Brongniart in the following particulars. "In that species the pinnæ succeed each other on the common rachis for a considerable space almost without alteration of form or size; but in *N. heterophylla* they diminish rapidly, and the pinnæ as well as the pinnules then assume very different forms. Towards the extremity of the pinnæ, on the contrary, the diminution is slower, the general figure more lanceolate, more acute, and the last lateral pinnules are much more lengthened. It is nevertheless possible that both plants may be only varieties of the same species;" a suspicion in which we are not only quite inclined to participate, but which we think might be well applied to many other cases.

DICTYOPHYLLUM CRASSINERVIUM.

So very little is known of the plants belonging to the new red sandstone formation, that during the progress of this work we have never before had an opportunity of describing a single species. For the fine specimen now presented to our readers we are indebted to J. Walker, Esq. President of the Liverpool Institution, who brought it to London at the request of Mr. Murchison.

It was found in what the latter gentleman considers the central part of the new red sandstone system of Great Britain, while excavating the Clarence Dock at Liverpool, and was presented to the Liverpool Institution by Dr. Traill.

The specimen is that of a leaf of considerable size, of which only a portion of the upper end remains, the end itself and all the margin being broken off. It bears a striking resemblance to the leaf of some of the thick-ribbed cabbages, consisting of several elevated ribs, full three-quarters of an inch wide, which spring at an acute angle from a midrib of about the same thickness, and divide towards the point into two or three branches, besides in one place putting out lateral ribs near the base. Intermediate to the principal ribs are, in one place, transverse connecting elevations which we may suppose to have been secondary veins; and in another place a small vein with lateral veinlets. In the whole specimen there is a good deal of irregularity of arrangement in the parts, and a greater want of symmetry than is usual in leaves.

Nothing like this has before been seen either from the new red sandstone, or any of the beds below the calk. With the Voltzias, and other fragments described by M. Adolphe Brongniart it has not a single character in common; and what is more important it is beyond all cavil a Dicotyledonous leaf. What it was there is, indeed, no evidence to shew; and there is no object in offering mere guesses upon the subject.

This is an interesting fact; for as it has been considered that our Dictyophyllum rugosum, t. 104, from the Oolite, is a Fern and not a Dicotyledonous leaf, as we however still believe, there was till now nothing positive to oppose to the opinion of a learned geologist that no Dicotyledonous remains, except of Coniferæ, would be found in the beds below the chalk. For ourselves we are persuaded that geology offers no ground for assuming the exclusion of Dicotyledons from the primitive flora; on the contrary, nothing opposed to the present design of the creation has yet been seen in any part of the flora of even the most remote periods, and if there is a general absence of the remains of Dicotyledons in the older rocks, it is surely more philosophical to ascribe that circumstance to the destruction of such plants, than to assume that this great class of vegetation which now comprehends two-thirds of all existing species, is of comparatively modern creation. To say this is little short of an assertion that the whole plan of organization of the Vegetable world has been altered since it was first called into existence out of chaos.

PECOPTERIS SERLII.

P. Serlii. *Ad. Brongn. Veg. Foss. p.* 294. *t.* 85.

From the Somersetshire Coal-field. It has been procured by M. Ad. Brongniart from the neighbourhood of Bath, from the coal quarry of St. Etienne, and also from Wilkesbarre in Pennsylvania.

Its remains are always very much broken; and yet the fragments of the leaves are well preserved; this would lead to the conjecture that it must have been of considerable size. Brongniart describes them as having been bipinnatifid; but they may have been much more compound. That they were of different shapes in different parts of the specimen is sufficiently shewn by the portion lying at the lower left hand corner of the plate, where the segments are not half the length of those on other parts of the specimen, and gradually diminish to the point.

LEPIDODENDRON STERNBERGII.

L. Sternbergii. *Suprà vol.* 1. *t*. 4.

We have already been so fortunate as to throw some light upon the nature of this extinct genus, in former articles in the present work. The accompanying plate will serve to complete its illustration as far as our knowledge at present goes.

It is the figure of a magnificent individual, which has lately made its appearance in the roof of the Bensham Coal seam, in Jarrow Colliery. It occurs in the great deposit of vegetable fossils which is there, as elsewhere, twelve to eighteen inches above the seam of coal, and fortunately has been overthrown or settled down in a north and south direction, which being that of the gallery, it is plainly to be seen throughout its whole length. Many much broader stems of Lepidodendra have occurred, but as they did not happen to lie in the direction of the workings, they could only be traced from side to side of the passages. This individual was thirty-nine feet in length, from the root to the first division into branches, the greatest breadth three feet, and gradually tapering; the ramifications of the branches extended to the coal on each side, the passage being $13\frac{1}{2}$ feet wide, and upwards of 10 feet in a direct line from the first division; here they were suddenly cut off, by a small dislocation, or slip of the strata. The specimen was pressed quite flat, and completely decorticated, having in its whole extent the general aspect of Sternbergs Lepidodendron rimosum (*tab.* 10. *fig.* 1.), which is well known to be the form of the internal surface of Lepidodendron. As before observed this is not by any means the largest stem of Lepidodendron which has occurred to us, but we figure it as being the most complete.

It is not we conceive an unfair mode of judging of the comparative magnitude of these extinct vegetables to compare the size of the rhomboidal scars with which they are covered, as it is obvious where such masses are concerned, we can seldom see them but in fragments. The unbroken stem of this individual was 39 feet long and 3 feet across, and the scars were not three-fourths of an inch in length; but in the Newcastle Museun there are several specimens beautifully preserved, where the markings are an inch and a quarter in length, and we lately observed on the coast of Northumberland, at a point a little south of Seaton Sluice, a mass of sandstone fallen from the cliff, where the scars were an inch and three-quarters in length; the surface of the mass was completely covered with the impression, which, although partly obliterated by the tide, was yet sufficiently distinct: it was seven feet long and four feet broad.

SPHENOPTERIS HÖNINGHAUSI.

Sphenopteris Höninghausi. *Ad. Brongn. Hist. des Végét. fossiles,* 1. 199. *t.* 52.

Originally sent to Adolphe Brongniart from Newcastle; our specimen is from Felling Colliery.

It forms one of a beautiful set of species, which the great describer of such plants refers to the genus *Cheilanthes.* It is known by its thick leaf-stalk, and oval or obovate, very obtuse, entire, or three-lobed pinnules.

SIGILLARIA FLEXUOSA.

From Killingworth Colliery, near Newcastle.

In many respects this is very like *S. reniformis*, from which, however, it differs in the presence of numerous, wavy, well defined lines, which apparently originate at the base of the leaves, and pass downwards, eventually losing themselves in the channels of the stem.

We have already expressed (*Vol.* 1. *p.* 153.) our opinion that Sigillarias are not the remains of Tree ferns. In a late number of his excellent *Histoire des Végétaux fossiles*, M. Adolphe Brongniart, defends with great acuteness and ingenuity the opinion that they were of that nature. To us, however, it still appears that the evidence is most defective and unsatisfactory, provided the genus Sigillaria is limited to the species with fluted stems: such as those which have been figured in this work But if it is made to comprehend all the scarred stems included in the genus by M. Brongniart, we in that case are quite ready to admit, that many (so called) Sigillarias, were either the rhizomata or trunks of Ferns.

206

LEPIDODENDRON OOCEPHALUM.

From Jarrow Colliery.

Apparently the fructification, in an incipient state, of a Lepidodendron allied to L. selaginoides and acerosum, but with more slender and longer leaves; the scale-like appearance upon the surface, is caused by the breaking off from their base of leaves similar to those shewn at the sides.

LEPIDODENDRON PLUMARIUM.

From Jarrow Colliery.

Whatever was the nature of the Lepidodendron gracile, figured at plate 9, the same it may be supposed was that of the species which forms the subject of the present figure. It consisted of a number of slender leaves, closely arranged over their stem, and curved upwards, like the plumes of a feather.

OTOPTERIS ACUMINATA;

var. BREVIFOLIA.

From the lower sandstone of Haiburn wyke near Scarborough.

Mr. Williamson, Jun., to whom we are indebted for the drawing, observes that the leaflets are considerably shorter and less acuminated than those of the first O. acuminata; they are also blunter. But these slight differences are no more than may be expected in different varieties of the same species.

SOLENITES ? FURCATA.

From the lower sandstone of Haiburn wyke near Scarborough; communicated by Mr. Williamson, Jun.

It occurs in the state here represented, the surface being marked by traces of delicate striæ. We place it in Solenites rather for the sake of giving the plant a station and a name, than because we have any reason for considering it of the same nature, further than its similarity of appearance.

210 A.

OTOPTERIS OVALIS.

From the oolitic formation of Gristhorpe Bay; communicated by Mr. Williamson, Jun. who says, that two specimens only have been discovered, one of which was much smaller than that figured. Both specimens had a small short petiole, but they were not found attached to any thing. They are composed of a thin brown substance which may be removed from the shale with a penknife. Each leaf consists of a midrib passing through to the point, and of veins either simple or once forked, planted perpendicularly upon it.

It is probable that the specimens are leaflets of some compound leaf, and possibly of a new Otopteris, to which genus we refer them.

CARPOLITHES ALATA.

Carpolithes alata. *Suprà vol.* 1. *t.* 87.

From Jarrow Colliery.

We presume this to be a more complete state of the fossil figured at plate 87 of the first volume of this book, and probably to be the same as the fig. 4. tab. 45. of Sternberg. In those specimens the shell has been broken and the interior laid open; here, on the contrary, we have a view of its external form, without however any additional information as to what it was.

ASTEROPHYLLITES RIGIDA.

From Jarrow Colliery.

This plant agrees very much with Asterophyllites comosa, from which it is distinguished by its stouter and more erect leaves, which have not, however, so well defined an outline as our engraver has given them, but are little more than long narrow stains upon the shale.

212

SPHENOPTERIS EXCELSA.

From the Newcastle Coal-field.

The specimens of this very beautiful fern are so imperfect, that we can neither ascertain what the margin was of the leaflets, nor the nature of their veins. It appears, however, to belong to the genus Sphenopteris in the neighbourhood of *Sp. Conwayi*, where it must stand until better evidence shall be produced as to its precise structure.

213

PECOPTERIS MARGINATA.

Pecopteris marginata, *Ad. Brongn. Hist. des Végét. foss.* 1. 291. *t.* 87.

A fern of frequent occurrence in the Coal of the North of England.

What is here represented is a pinna of a bipinnatifid leaf; the pinnules seem to have been drawn a little together before the plant was fixed in the matrix, and there is every appearance of its having been some hard-leaved species.

M. Ad. Brongniart compares it with Pteris biaurita.

SPHENOPTERIS CUNEOLATA.

From the Newcastle Coal-field.

The forked stem of this fern approximates it to S. artemisiæfolia and its allies, from which it is distinguished by the narrow, wedge-shaped, entire, or emarginate lobes of its leaves. In that respect it may be compared with Sp. furcata, but that plant has a regularly bipinnatifid leaf.

Not a trace of veins could be found in the specimen from which the drawing was made.

215

PECOPTERIS OREOPTERIDIS.

Pecopteris oreopteridis. *Ad. Brongn. Hist. des Végét. fossiles,* 1. 317. *t.* 105. *f.* 1, 2, 3.

One of the finest of the coal-measure ferns, and in all probability the remains of an arborescent species. Sent from Welbatch, near Shrewsbury, by the Rev. W. Corbett.

Its bipinnate character and its large size will prevent its being easily mistaken for any thing except *P. polymorpha,* and from that species M. Ad. Brongniart well distinguishes it by the veins of the leaflets, which, here, are simple or only forked, and in the former plant dichotomous or twice forked.

CALAMITES APPROXIMATUS.

Calamites approximatus. *Adolphe Brongniart Hist.* 1. 133. *t.* 24. *t.* 15. *f.* 7, 8.—*Artis Antediluv. phytol. t.* 4.

The accompanying drawing was made from a beautiful specimen of this plant, nearly fifteen inches long and three inches and three-quarters wide, communicated by Professor Buckland from Camerton.

It agrees in a striking manner with the figures of Artis and Adolphe Brongniart, with the addition of a number of pits placed on the articulations, in a quincunical manner, as in *Calamites cruciatus*. Hence it is probable that the latter supposed species will require to be reduced to *C. approximatus*.

CYCLOPTERIS OBLATA.

A noble species of this curious genus, sent us by Matthew Dawes, Esq. of Acresfield, Bolton-le-Moor, from Little Lever near the latter place.

It is obviously different from all the species hitherto discovered, not only in its unusual size, but also in the peculiar oblate form of the leaf. The specimen appears to have been extremely delicate, for it is much puckered and plaited across the veins, as is happily expressed by our artist in his figure.

Probably *Adiantites giganteus* of Göppert is something of this sort.

BOTHRODENDRON PUNCTATUM.

Bothrodendron punctatum. *Supra vol.* 2. *tab.* 80.

It is not uncommon to find in the sandstone of the coal measures such bodies as that now represented: oblique flattened convexities, having a circular scar or two at their apex, and their sides indistinctly marked by depressed or elevated lines, losing themselves in the sandstone.

These bodies are the bases of the cones of Bothrodendron, and are already alluded to at Plate 80. To what is there said concerning them we find nothing to add. Such a cone as the one before us once fitted into the sockets shewn on the stem at Plate 80; its apparent apex is the base, its true apex is lost in the sandstone, and the scar at the centre of their convexity is where they joined the trunk.

The specimen here represented was communicated by D. D. G. Lloyd, Esq. from Ketley, east of the Wrekin, a freestone quarry belonging to Lord Gower.

BRACHYPHYLLUM MAMMILLARE.

Brachyphyllum mammillare. *Suprà tab.* 188.

We have been induced to republish this plant in the hope that a second figure would convey a more correct idea of it than the first, in which the ends of the branches were represented too thick, and the points of the leaves in the magnified view too sharp, convex, and recurved. In some respects the present plate is more correct, but the magnified portion is made to resemble too much the surface of a Lepidendron, covered with lozenge-shaped scars.

In reality the fossil has its old branches closely covered with short, ovate, rather obtuse, appressed, ribless, scale-like leaves, which diminish in number as the branches diminish, till, without altering in form, they become on the youngest twigs merely alternate. It should be compared with such recent Coniferous plants as are represented at t. 127 of this work.

CARPOLITHES SULCATA.

In ironstone on Wardie Beach, near Newhaven, found by the Lord Greenock.

This fruit may be thus described. Ovate, three-eighths of an inch wide, by five-eighths long, chalky white, tapering to the point, very slightly depressed at the base, with about ten deep furrows, which do not reach the base, apparently one-celled, with a thick pericarp.

TRIGONOCARPUM DAWESII.

From Peel stone quarry near Bolton. Mr. Dawes.

Oblong, rather wider at one end. Two inches and a quarter long, an inch and a quarter wide, with three slight nearly equidistant angles. Probably the nut of a Palm.

2. & 4. TRIGONOCARPUM NÖGGERATHI.

See Plate 193. B.

---

1. & 3. TRIGONOCARPUM OLIVÆFORME.

From Peel stone quarry, near Bolton. Mr. Dawes.

These are obviously Palm fruits (*see vol.* 2. *plate* 142.) T. olivæforme has only three angles instead of six, and is more ovate; otherwise it hardly differs from T. Nöggerathi.

PECOPTERIS BUCKLANDII.

Pecopteris Bucklandii. *Ad. Brongn. Hist. des Végét. fossiles* 1. 319. *t.* 99. *f.* 2.

From the Newcastle coal-field.

Our specimen is in the same state as that sent from Camerton to M. Brongniart by Dr. Buckland; so that we are equally unable to say whether it is a species with a tripinnatifid or bipinnatifid leaf. It is, however, easily known by its small size, as compared with P. oreopteridis, &c. to which it approaches. M. Brongniart says that the very oblique direction of the veins, which are bifurcate near their base, and sometimes forked again near the top of their upper arm, constitutes a good distinguishing character.

224 and 225

STERNBERGIA APPROXIMATA.

Sternbergia angulosa. *Artis Antedil. phytol. t.* 8.
Sternbergia approximata. *Ad. Brongn. Prodr. p.* 137.

A most singular coal-measure plant, occurring in most of the coal-fields of this country, but not abundant any where. The specimens are usually found in Sandstone, and are covered with fine coal, which either adheres in the form of an even thick glossy integument, or adheres in a powdery state to the surface of the stem.

The specimens figured in plate 225, are from Somersetshire and Newcastle; that in plate 224 is from Halliwell stone quarry near Bolton, from Mr. Dawes. They differ in size, in being more or less angular, and in the distance between their cross bars; but we see nothing to justify us in considering such characters as of specific importance.

When the integument of coal is broken off, these plants are sometimes found simply marked by horizontal depressed lines, which meet alternately from opposite sides anastomozing in the middle; but in other cases the space between the lines is excavated into deep furrows, and honey-combed as it were by the formation of short perpendicular bars which connect the lines; traces also may be found of lines running along the sides of the stem for a considerable distance. The result of this is that many stems appear as if they were composed of horizontal plates, about 1-16th of an inch apart and held together by some connection in the axis of the stem: a most extraordinary appearance, to which we know of no parallel, and which we are by no means prepared to say is their real structure.

M. Adolphe Brongniart regards such stems as analogous to those of Yucca or Dracæna, considering the horizontal lines as the stations of leaves which have fallen off. We regret to say that we have no evidence to produce either in confirmation or refutation of this opinion, beyond what the plates and the above remarks afford.

226 A

ZAMIA OVATA.

We received this beautiful and unique specimen from W. Richardson, Esq., F.G.S. who found it upon the coast of Kent, near Feversham. The surface of the land in and about that place is covered with bouldered green sand fossils, to which this also belongs.

It is evidently of the same nature as the cone figured at plate 125 of this work, but appears to be distinguished specifically by the ends of the scales being acute or nearly so, and not truncate, and by the ovate form of the cone itself. The specimen is much rolled; but whether these differences can be assigned to that circumstance must be shewn by further evidence.

It is not a little curious that the only two Zamia cones yet found in the green sand should both have been partially devoured on one side. Here the figure A. 2. shews that the whole interior of the cone is laid open on one side, as in Professor Henslow's specimen of Zamia macrocephala.

The cone is broken off at the bottom, and was therefore longer than what is figured: but it is probable that the fracture has taken place near the base.

STROBILITES WOODWARDI.

Communicated by Mr. Woodward of Norwich with the following note.

"The strobilus, originally three inches long, was found in a layer of decayed wood and vegetable matter at the base of Paston Hill, near Mundesley, a portion of the line of the Norfolk cliffs. It is seen in numerous other places in the cliffs between Hasboro and Cromer, and consists of wood, bark, leaves, seed-vessels, &c. an accumulation similar to those seen on the surface of a plantation. I consider this layer to be the remains of the forest trees of the antediluvian period, as, above it, are in some places 10 feet, in others more than one hundred feet, of diluvial debris. And although there may be some similarity in the appearance between these remains and those found in the forming of the Dilham canal, about four miles inland from the line of these cliffs, still it is evident from the nature of the remains in that locality that they are Lacustrine, and of the post-diluvian period."

The cone is in the state of half charred wood, or rather in that of wood met with in the brown coal formation; and has a strong woody centre, round which are loosely arranged a considerable number of woody scales as in the cone of the European Larch. The scales are so much broken that their original form cannot be ascertained with certainty; it is probable, however, that they were entire, rounded, rather oblate, and thinned off to the edge, as in the genus Larix.

Possibly this may be referable to the genus Voltzia, of which all the species hitherto discovered are from the new Red Sandstone, or *grès bigarré.*

227 A

ENDOGENITES STRIATA.

We sometimes find in the coal-measures bits of a stem with a slight carbonaceous coating, below which pass numerous parallel furrows. It is to be supposed that the furrows are the woody fibres of an Endogen, and the carbonaceous coating its cortical integument, which stands in room of bark.

227 B

CARPOLITHES AREOLATA.

From the oolite of Scarborough. Professor Buckland.

A conical angular body, divided into irregular areolations, with prominences here and there which bear no definite relation to such areolations, and over which it seems as if a coating of cellular substance had been drawn; such is all that can be made out in this production.

HALONIA REGULARIS.

Fig. 1. from Halliwell stone quarry near Bolton; fig. 2. from Peel stone quarry near Bolton; both communicated by Mr. Dawes.

These are most remarkable specimens of this curious genus. They are quite distinct both in dimensions, and in the regularity with which their tubercles are arranged, from either of the species previously figured.

FILICITES SCOLOPENDRIOIDES.

Filicites scolopendrioides. *Ad. Brongn. Ann. des sc. nat. xvi.* 443. *t.* 18. *f.* 2. *Histoire des veg. foss.* 1. 388. *t* 137. *f.* 2.

The specimen figured by M. Adolphe Brongniart under this name from the grès þigarré of Sultz-les-Bains is in a state of fructification. That from which our plate has been taken was barren; we received it from Mr. J. S. Bowerbank, who procured it from the new Red Sandstone near Whitby. It was so long that we can only shew the upper and lower extremities, and is chiefly interesting as proving that the principal distinction between the barren and fertile pinnæ of the plant consists in the former being adherent to the rachis by their whole base, slightly falcate and obtuse, as is represented in M. Brongniart's figure at the bottom.

No evidence concerning the veins was afforded by Mr. Bowerbank's specimen; but it proved most distinctly that the leaf was not simple as M. Brongniart supposes, with parallel simple oblique rows of fructification, but pinnated, with the upper pinnæ covered all over their back with fructification. The barren pinnæ either occupied only the lower part of the leaf, or, as in the specimen before us, constituted the whole leaf, as in the modern Acrosticha aureum and inæquale.

We consider this plant nearly allied to the Indian Acrostichum Wightianum.

SPHENOPTERIS LINEARIS.

Sphenopteris linearis. *Sternb. Fl. t.* 42. *f.* 4. *Ad. Brongn. Hist. des Végét. fossiles,* 1. 175. *t.* 54. *fig.* 1.

Found by Count Sternberg in the Bohemian Coal-field, and by Dr. Hibbert and others in that of the North of England.

The broad wedge-shaped pinnules, with truncated lobes, which have pretty generally two or three shallow crenatures at their end, together with a compactness of growth not common in the genus, mark this species at first sight.

The plant figured at plate 181 under the name of *Sp. furcata* is too near this. The true *Sp. furcata* is more divaricated with longer and narrower lobes to its pinnules.

INDEX TO VOLUME III.

END OF VOLUME III.

NORMAN AND SKEEN, PRINTERS, MAIDEN LANE, COVENT GARDEN.

THE

FOSSIL FLORA

OF

GREAT BRITAIN;

OR,

FIGURES AND DESCRIPTIONS

OF THE

VEGETABLE REMAINS FOUND IN A FOSSIL STATE

IN THIS COUNTRY.

BY

JOHN LINDLEY, Ph. D. F.R.S. &c.

PROFESSOR OF BOTANY IN THE UNIVERSITY OF LONDON;

AND

WILLIAM HUTTON, F.G.S. &c.

"Avant de donner un libre cours à notre imagination, il est essentiel de rassembler un plus grand nombre de faits incontestables, dont les conséquences puissent se déduire d'elles-mêmes."—*Sternberg.*

PART I. OF VOLUME III.

LONDON:

JAMES RIDGWAY AND SONS, PICCADILLY.

MDCCCXXXVII.

m. del.t

Pub by Mess.rs Ridgway. London. July 1835.

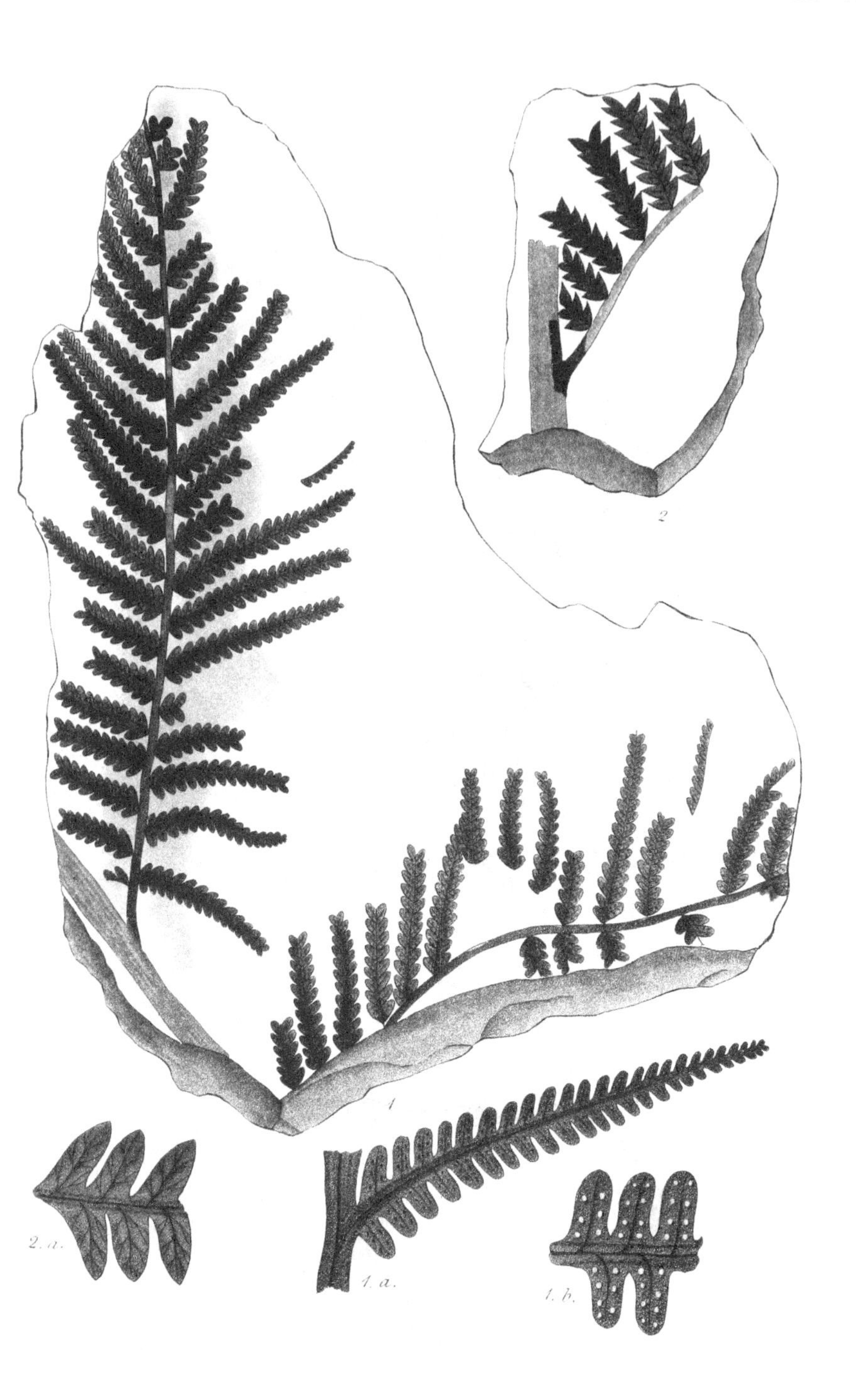

Pub: by Mefs.rs Ridgway, London, July, 1835.

Natural Size.

Pub by Mefs.rs Ridgway. London. July. 1836.

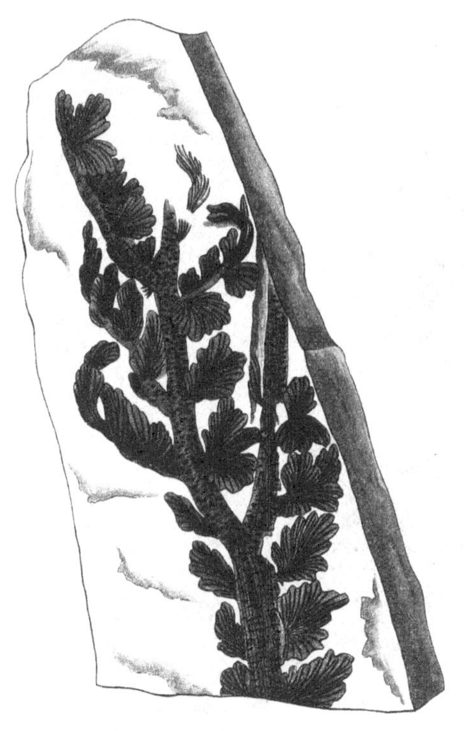

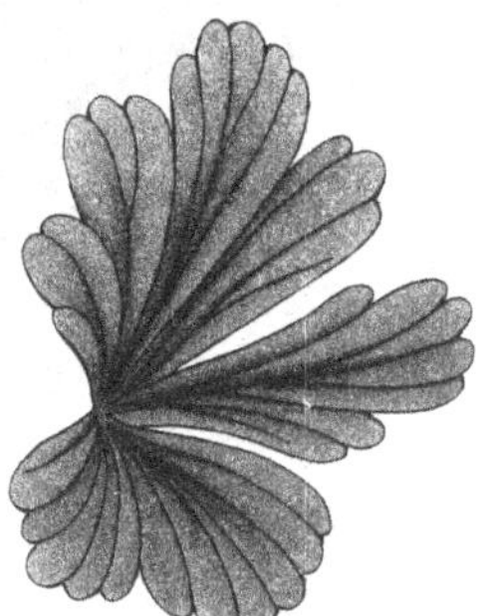

Magnified.

Pub: by Mefs.rs Ridgway, London, July, 1835.

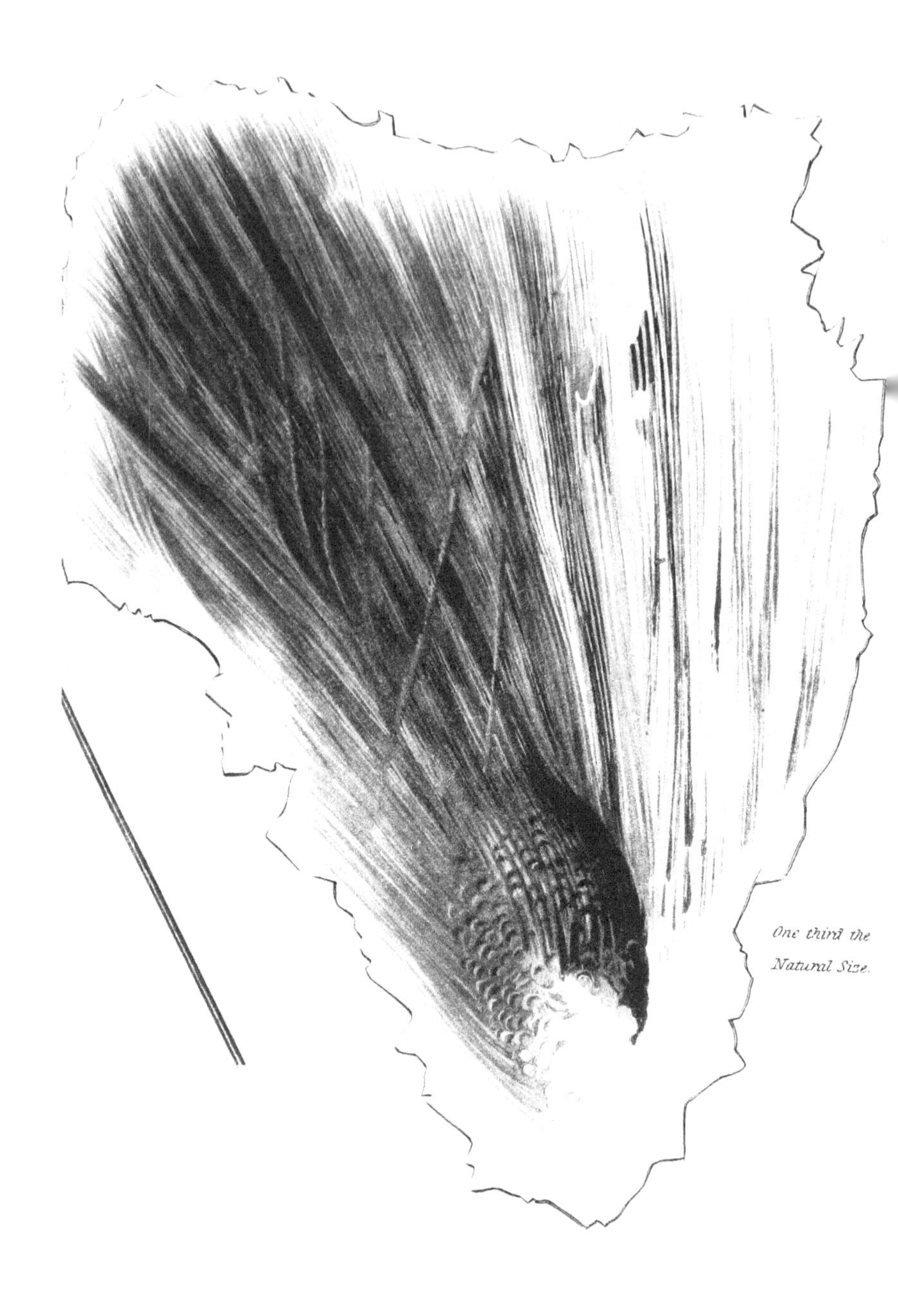

One third the Natural Size.

Pub: by Mess.rs Ridgway. London . July 1835.

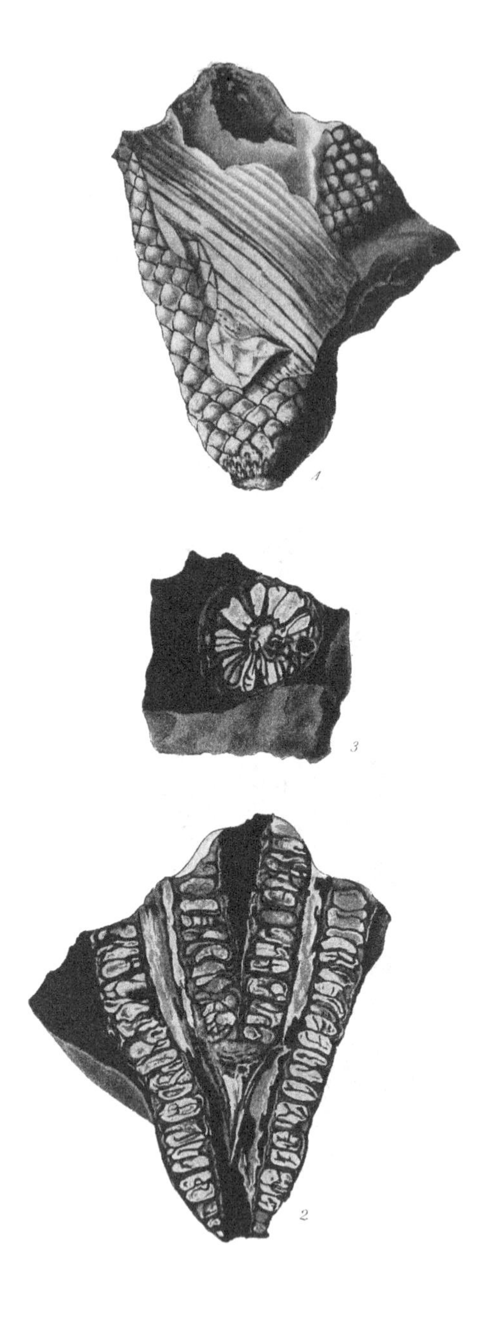

Pub by Mefs.rs Ridgway, London, July 1835.

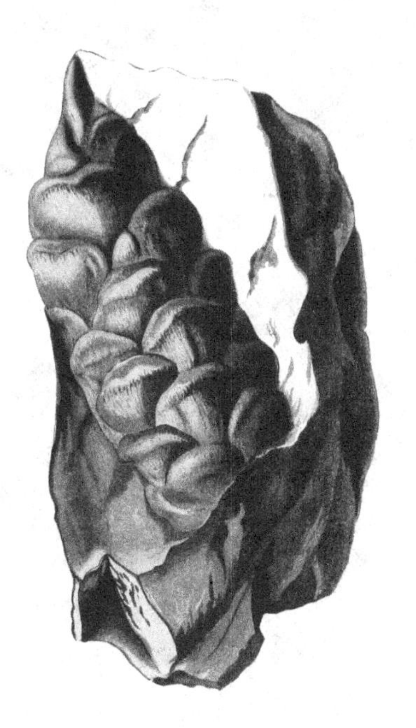

Pub. by Mess.rs Ridgway, London. July 1835.

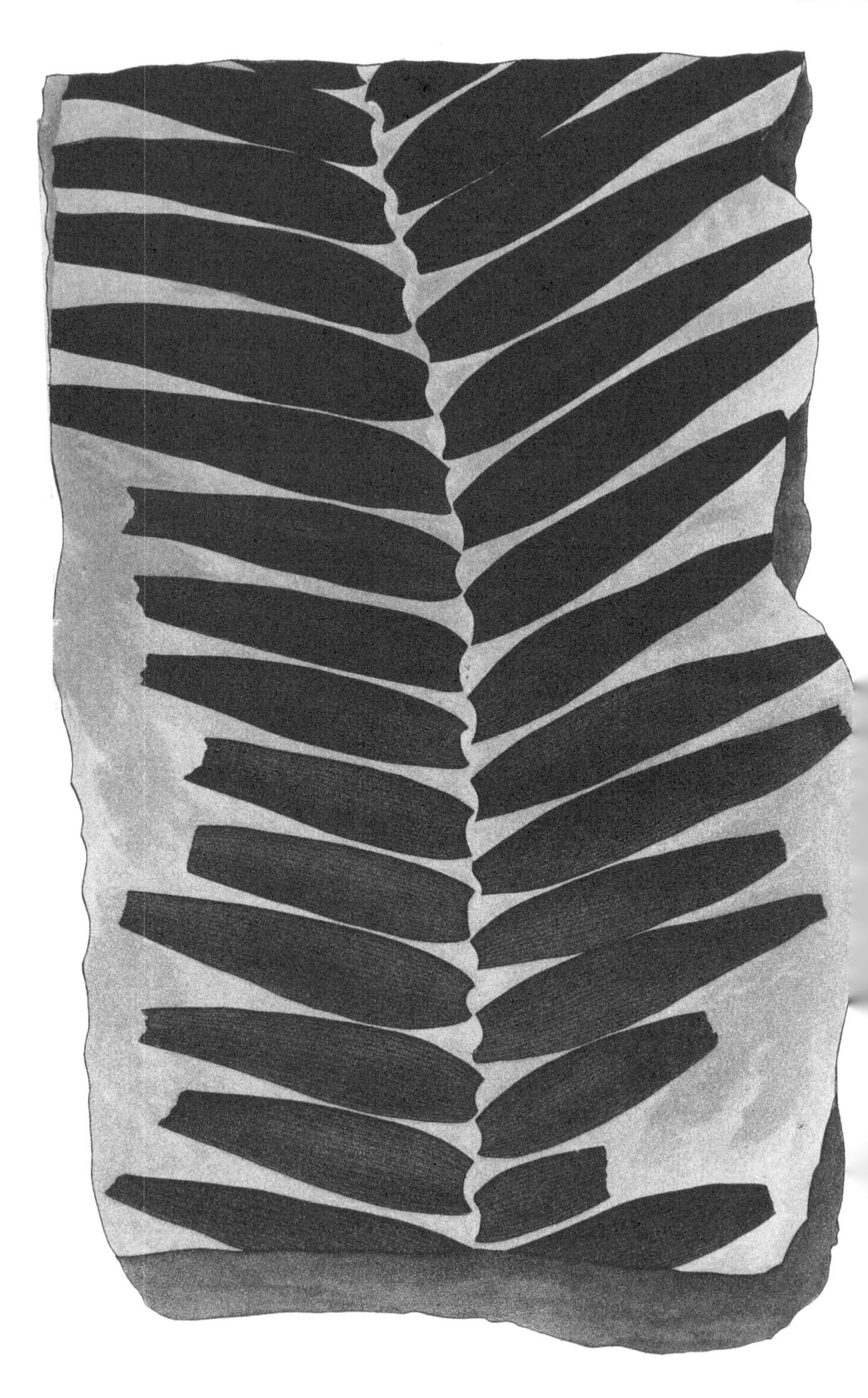

Pub: by Messrs Ridgway. London. July. 1835.

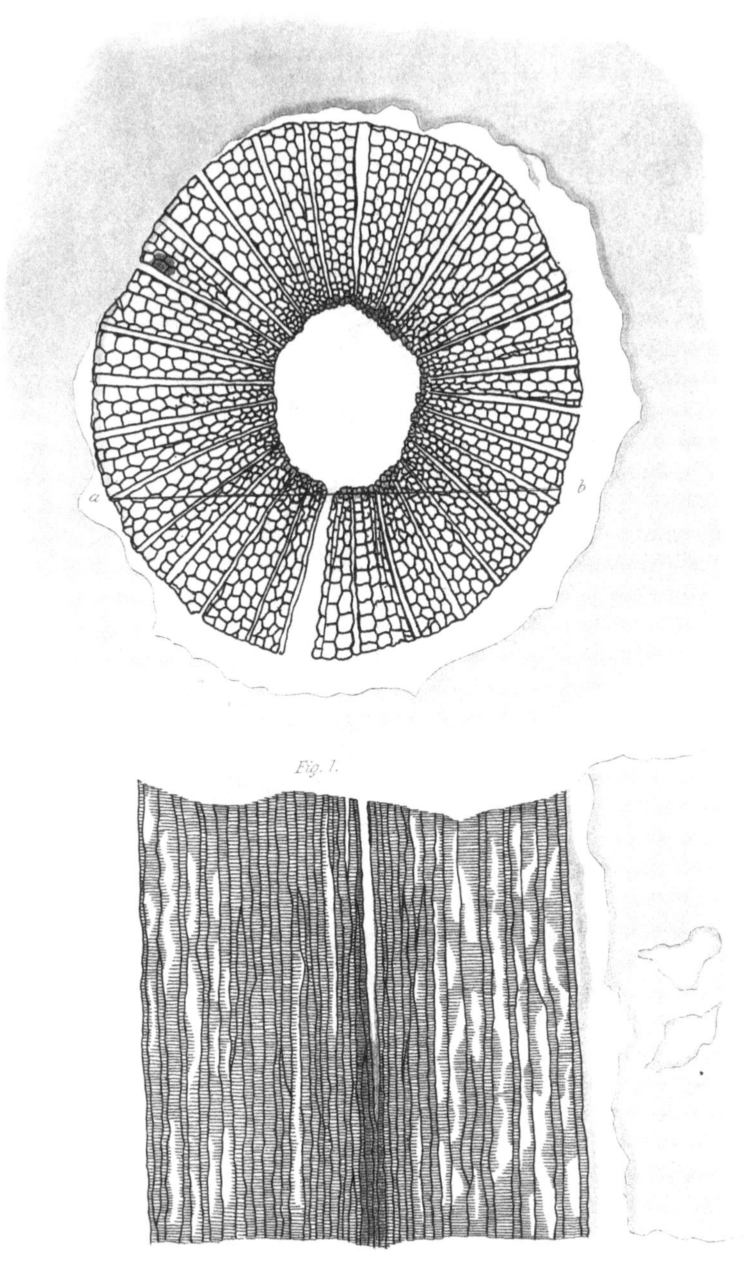

Fig. 1.

Fig. 2.

Pub: by Messrs Ridgway. London. July. 1835.

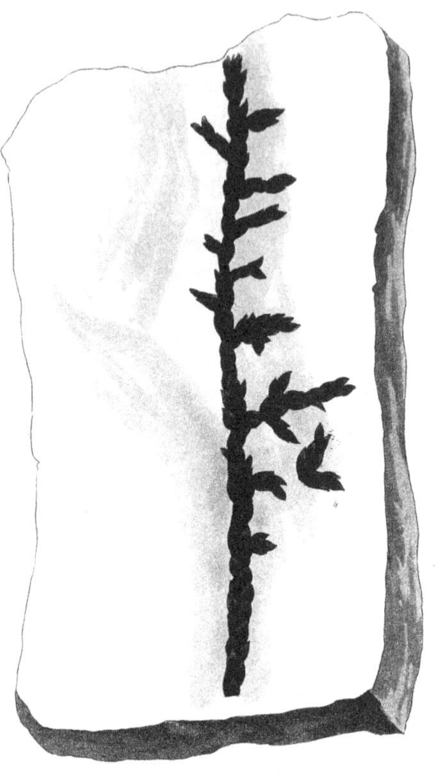

Magnified.

Pub. by Mefs.rs Ridgway, London, Oct. 1835.

Magnified.

Pub by Mess.rs Ridgway. London. Oct. 1835.

Pub. by Mess.rs Ridgway. London. Oct. 1835.

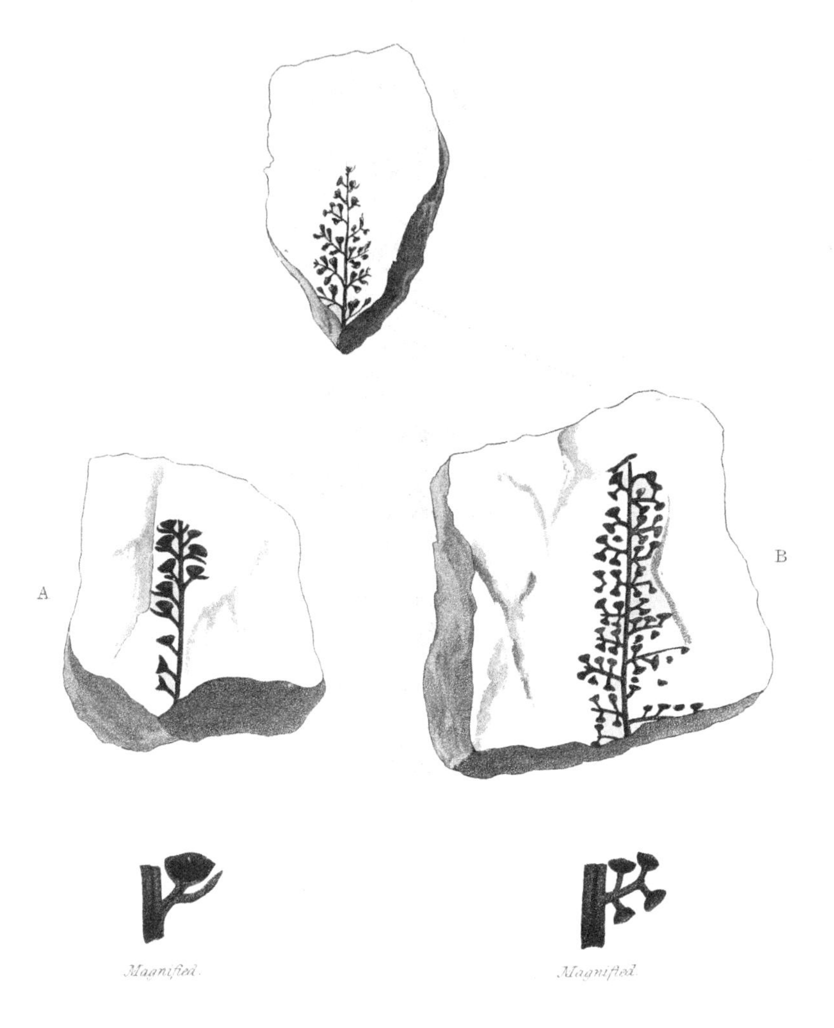

Pub: by Mess.rs Ridgway. London. Oct. 1835.

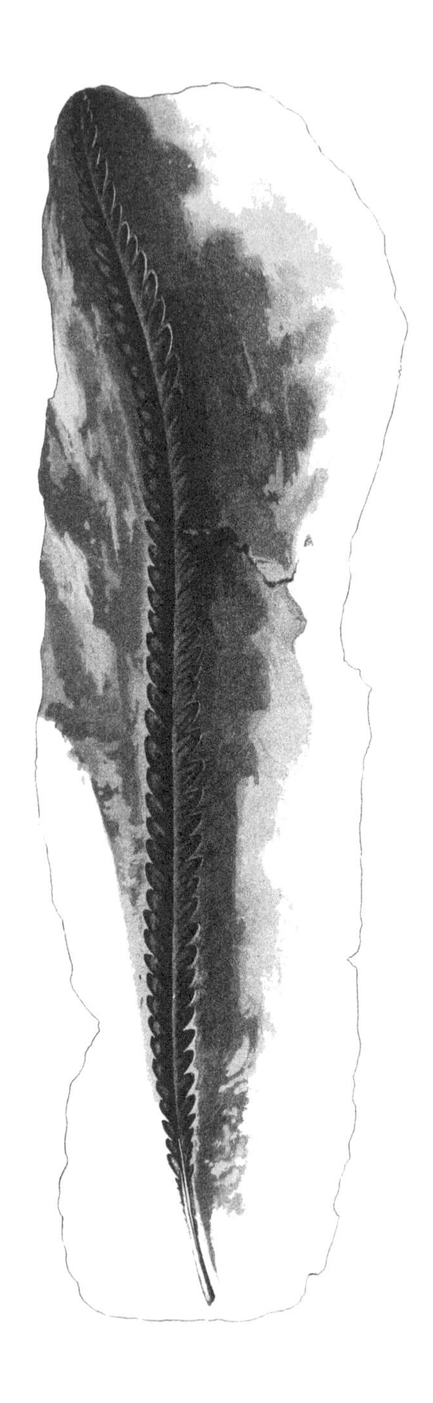

Pub: by Mess.rs Ridgway. London. Oct 1835.

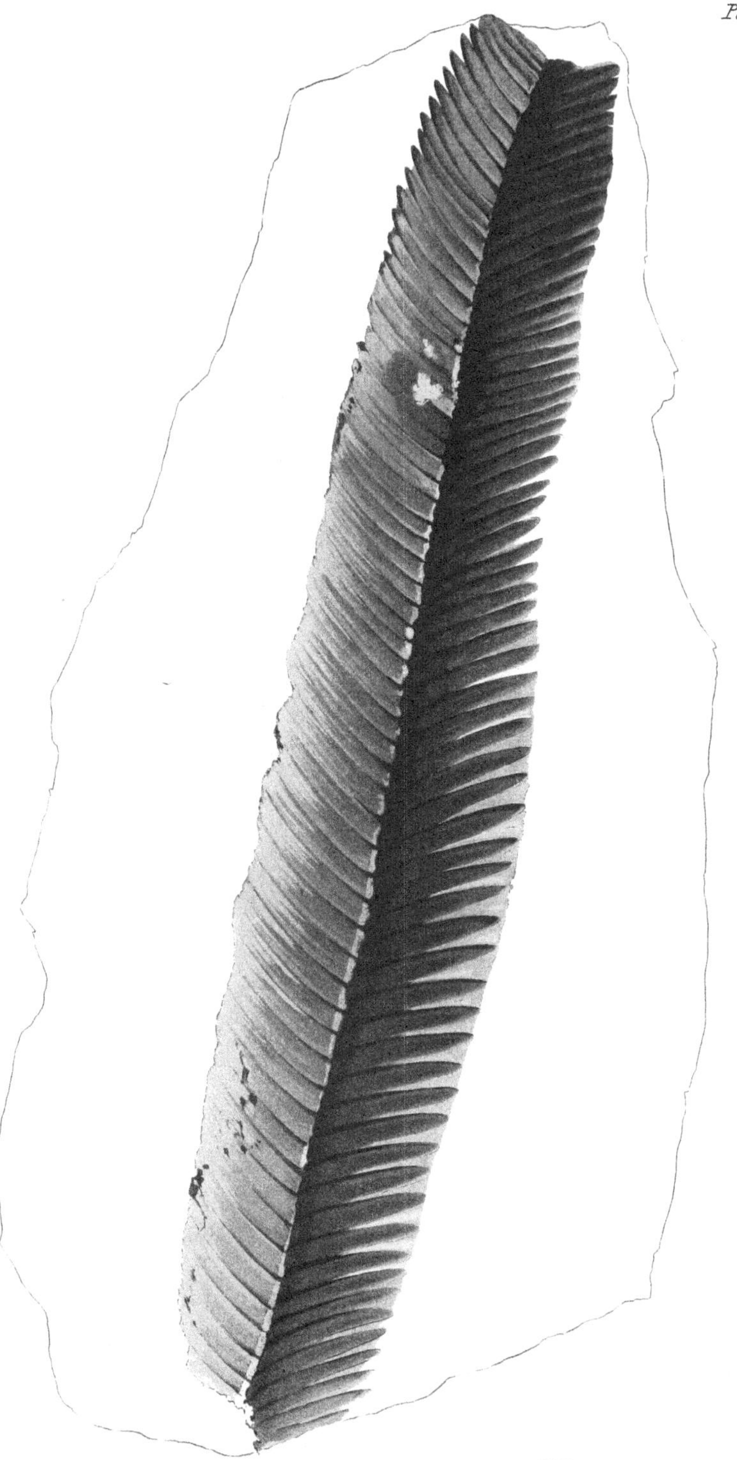

Pub. by Mess.rs Ridgway London, Oct. 1835.

Pub. by Mefs.rs Ridgway. London. Oct. 1835.

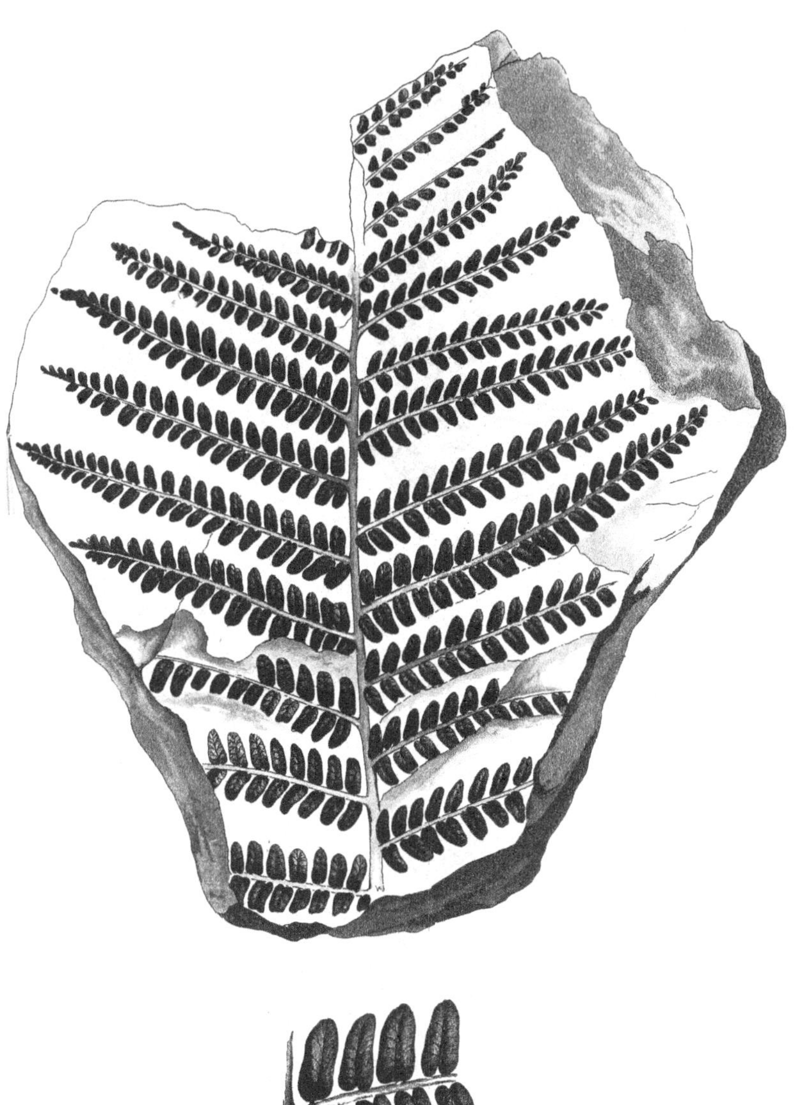

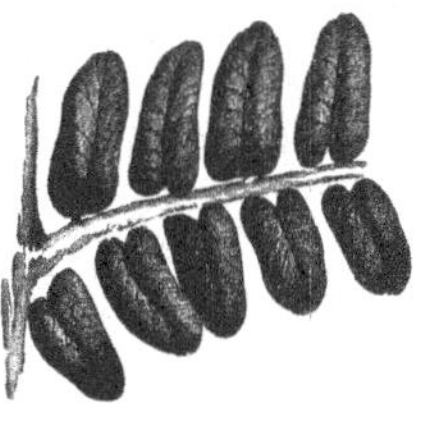

Magnified.

Pub: by Mess.rs Ridgway. London. Oct 1835.

Pub: by Mefs.rs Ridgway, London, Oct. 1836.

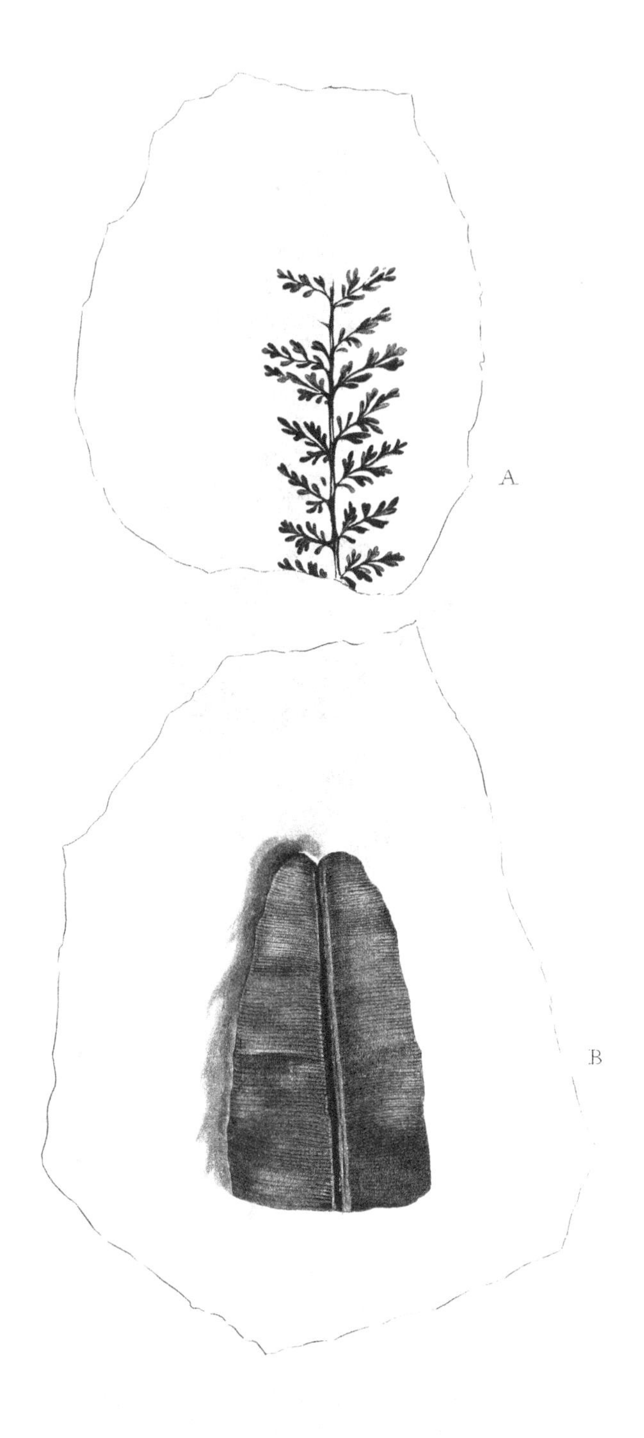

Pub: by Mefs.rs Ridgway London Oct. 1836

Pub. by Mess.rs Ridgway, London, Jan.y 1836

The material originally positioned here is too large for reproduction in this reissue. A PDF can be downloaded from the web address given on page iv of this book, by clicking on 'Resources Available'.

Plate 178.

Magnified.

Pub: by Mess.rs Ridgway, London, Jan.y 1836.

Magnified.

Pub: by Mefs.rs Ridgway, London, Jany 1836.

Pub: by Mefs.rs Ridgway, London, Jan.y 1836.

Pub. by Mefs.rs Ridgway, London, Jany 1836.

Pub. by Mefs.rs Ridgway, London, Jan.y 1836.

Pub by Mefs^rs Ridgway, London, Jany 1836.

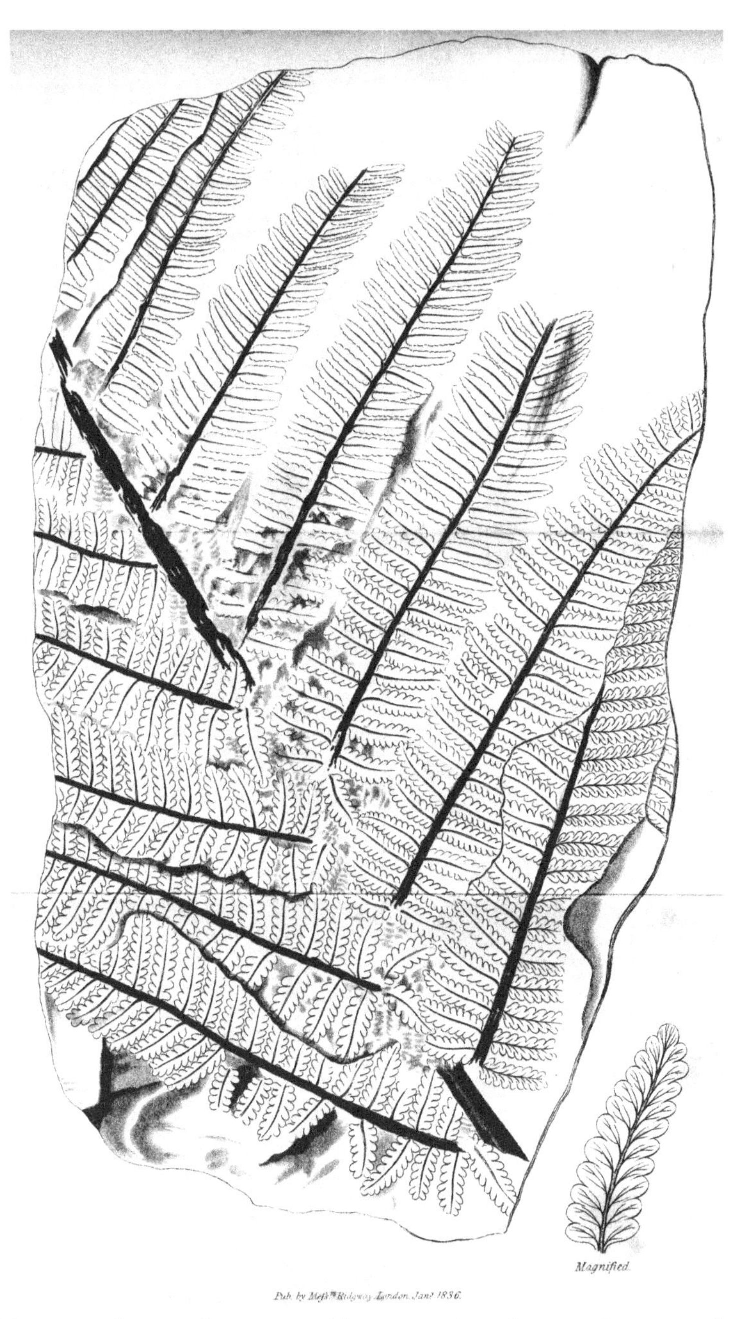

Pub. by Mess.rs Ridgway London Jan.y 1836.

The material originally positioned here is too large for reproduction in this reissue. A PDF can be downloaded from the web address given on page iv of this book, by clicking on 'Resources Available'.

Pub: by Mess.rs Ridgway, London. Apr 1836.

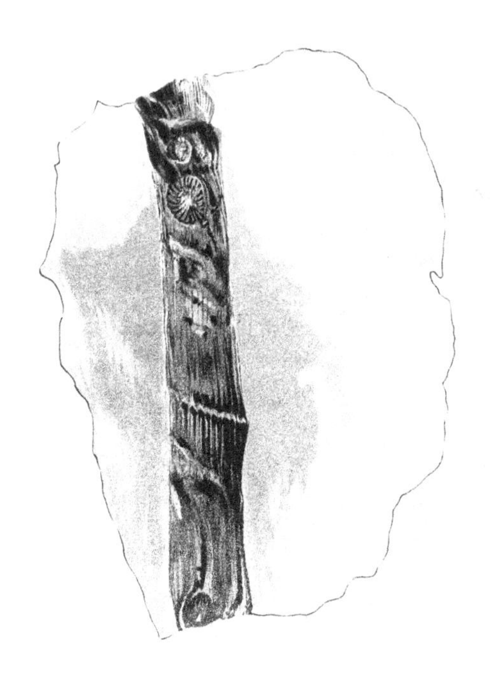

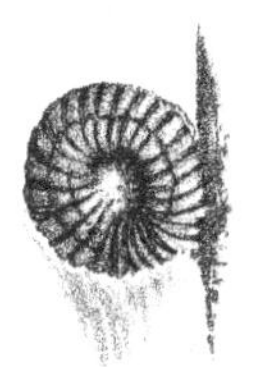

Magnified

Pub. by Mefs.rs Ridgway, London, Apr 1836.

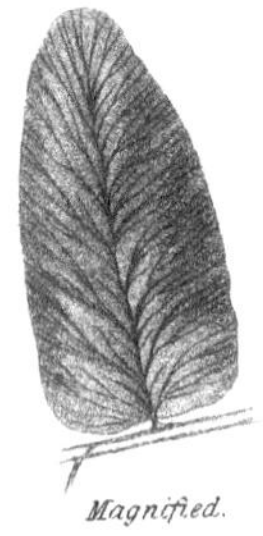

Magnified.

Pub: by Messrs Ridgway, London, Apr. 1836.

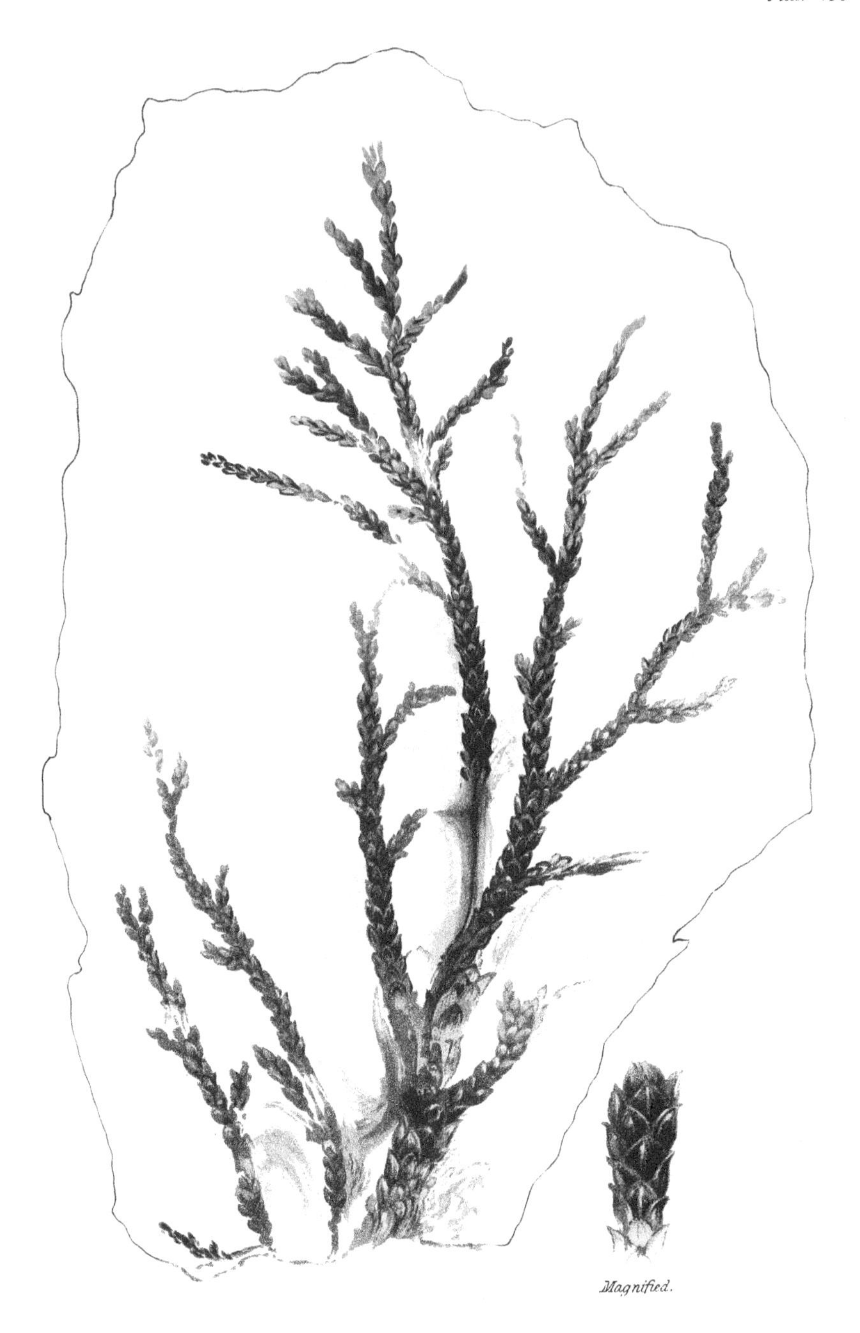

Pub: by Mess.rs Ridgway, London. Apr. 1836.

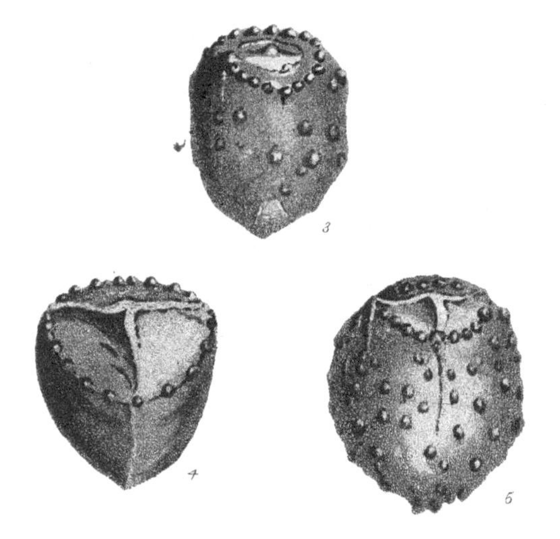

Pub: by Mess.rs Ridgway, London, Apr. 1836.

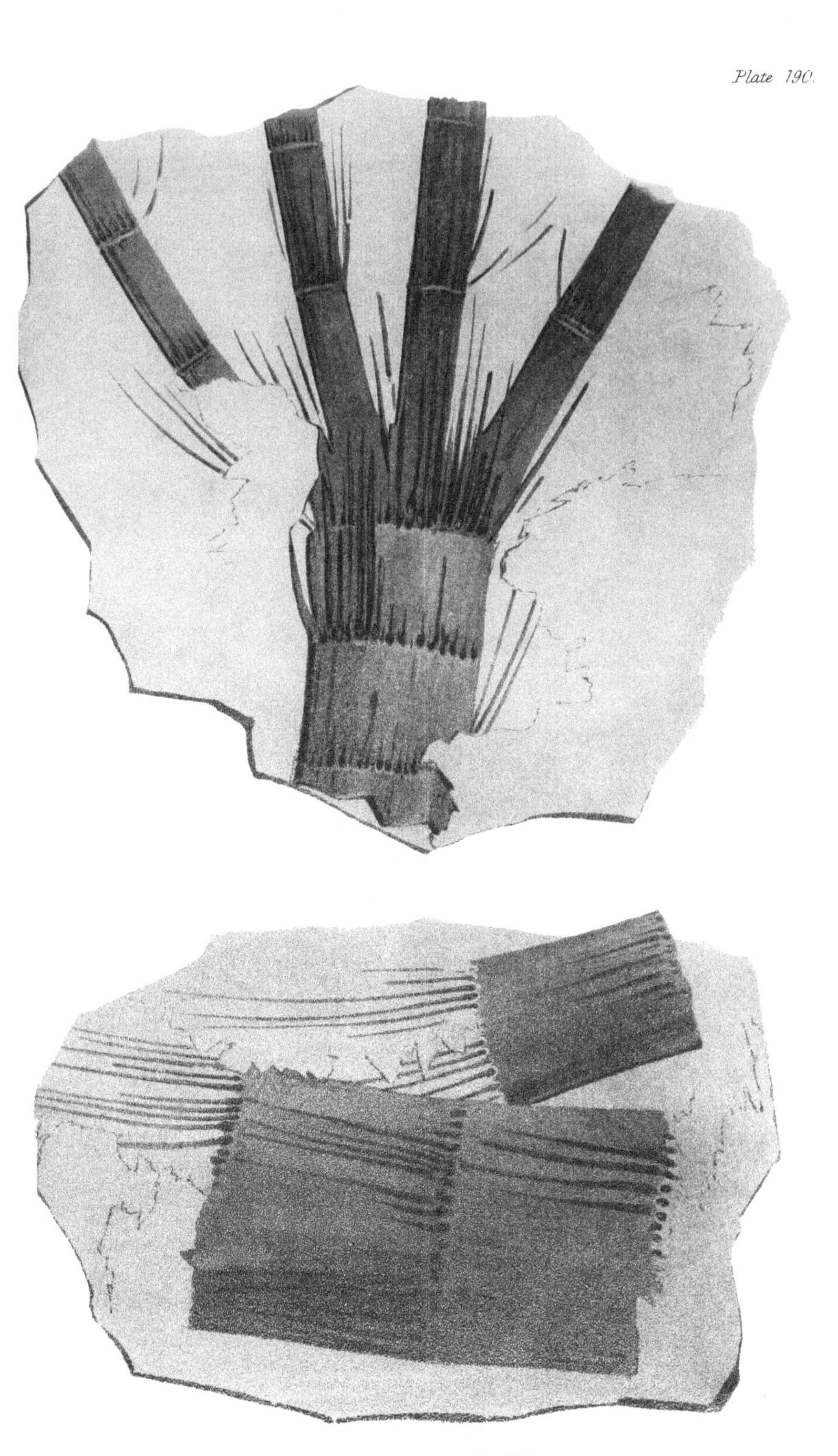

Pub: by Mess.rs Ridgway, London. Apr. 1836.

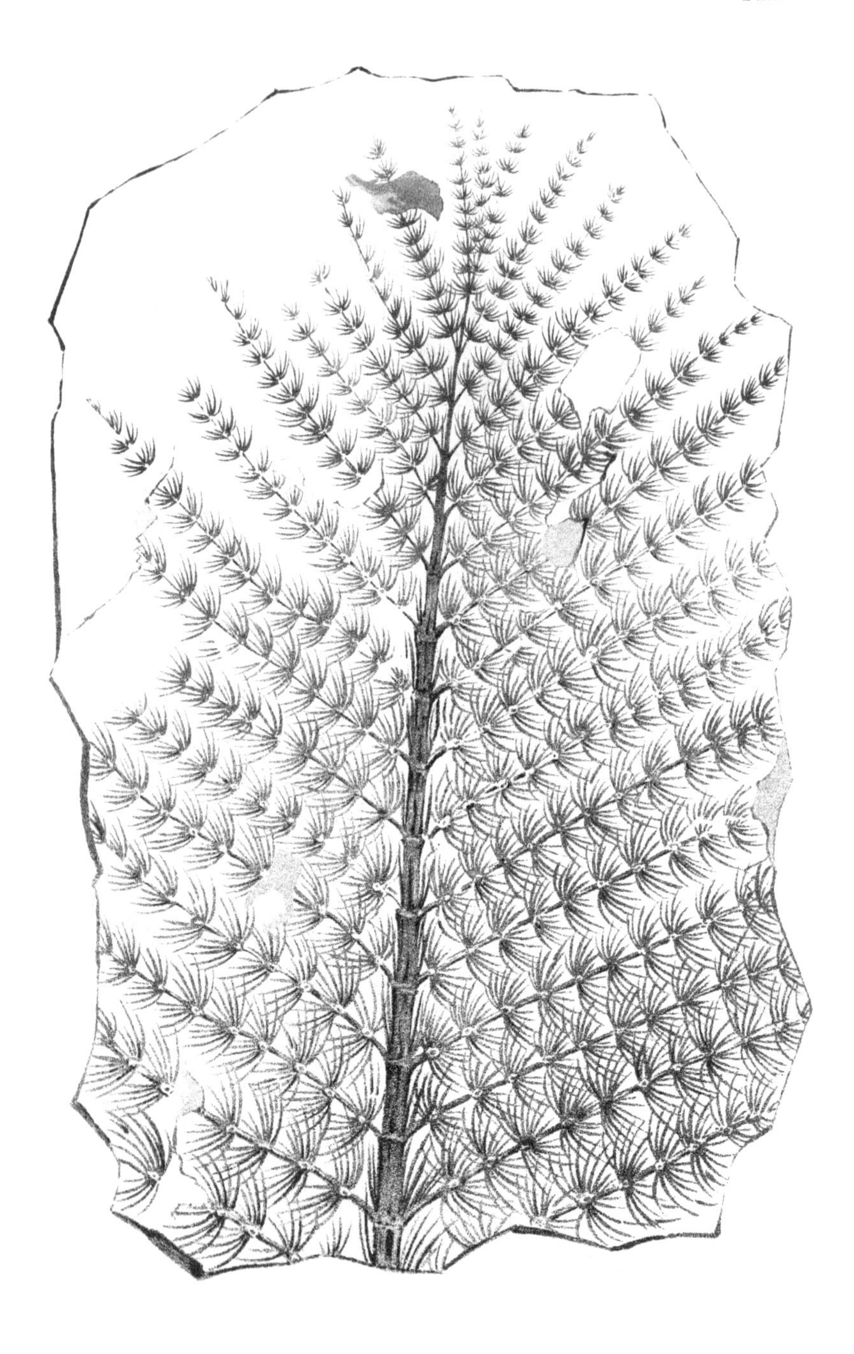

Pub: by Mess.rs Ridgway, London, Apr. 1836.

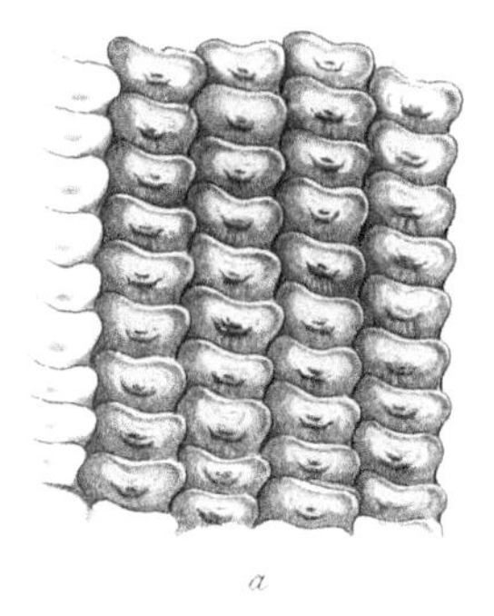

a

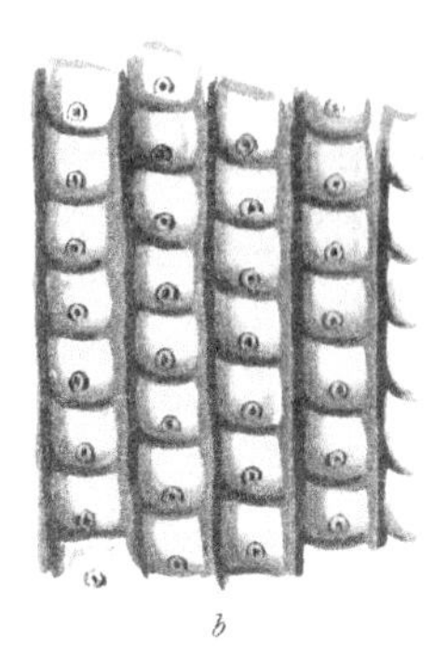

b

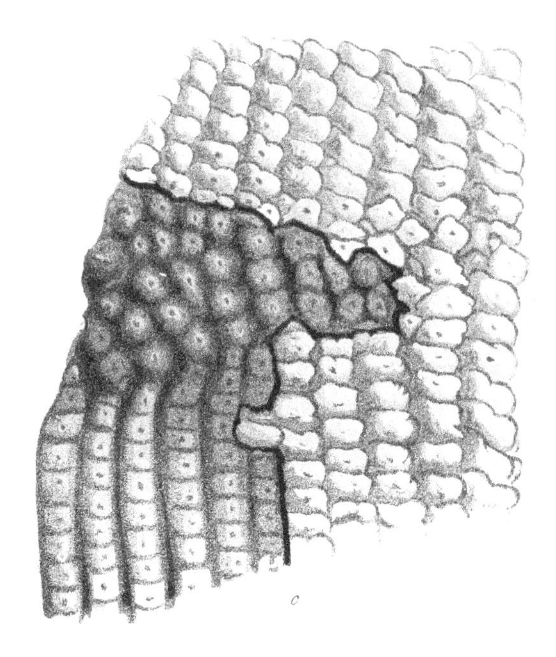

c

Pub. by Mess.rs Ridgway, London, Apr. 1836.

Pub. by Mess.rs Ridgway, London, Apr 1836

Pub. by Mess.rs Ridgway, London, Apr. 1836.

Pub. by Mefs.rs Ridgway Jan.y 1837.

Pub. by Mess.rs Ridgway Jan.y 1837.

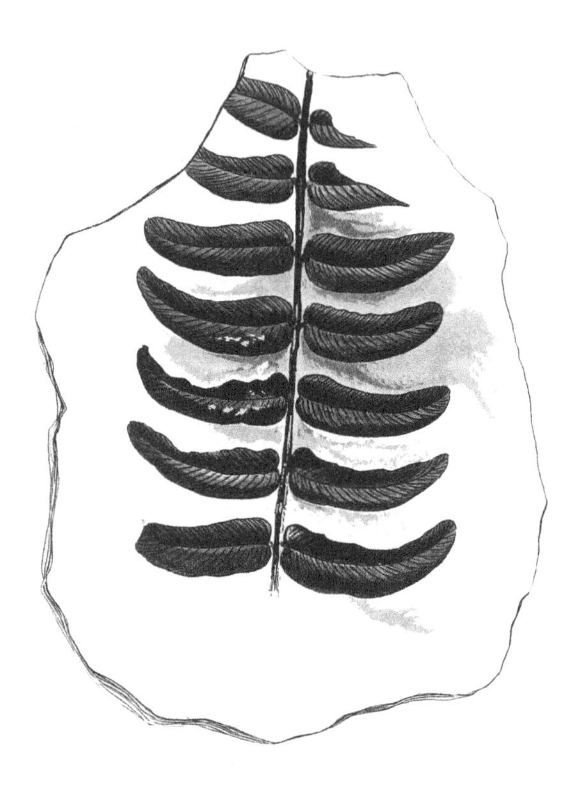

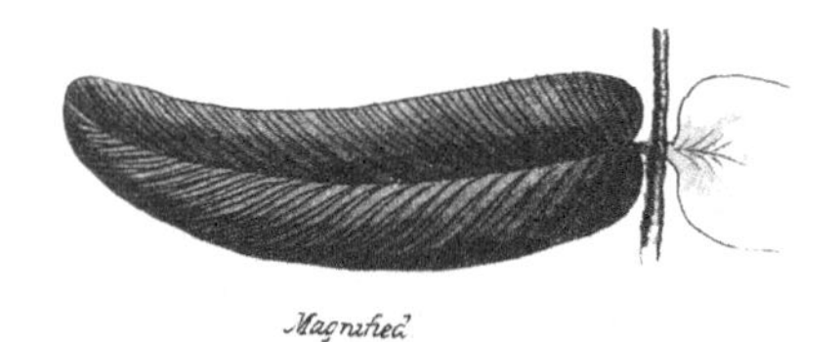

Magnified.

Pub by Messrs Ridgway, London, Jany 1837

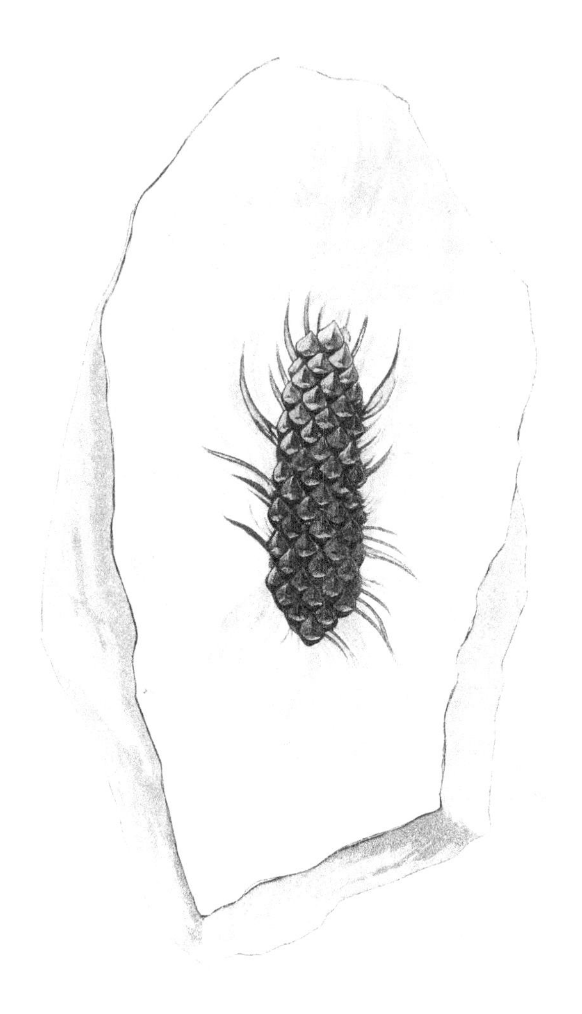

Pub. by Messrs. Ridgway London, Jan.y 1837

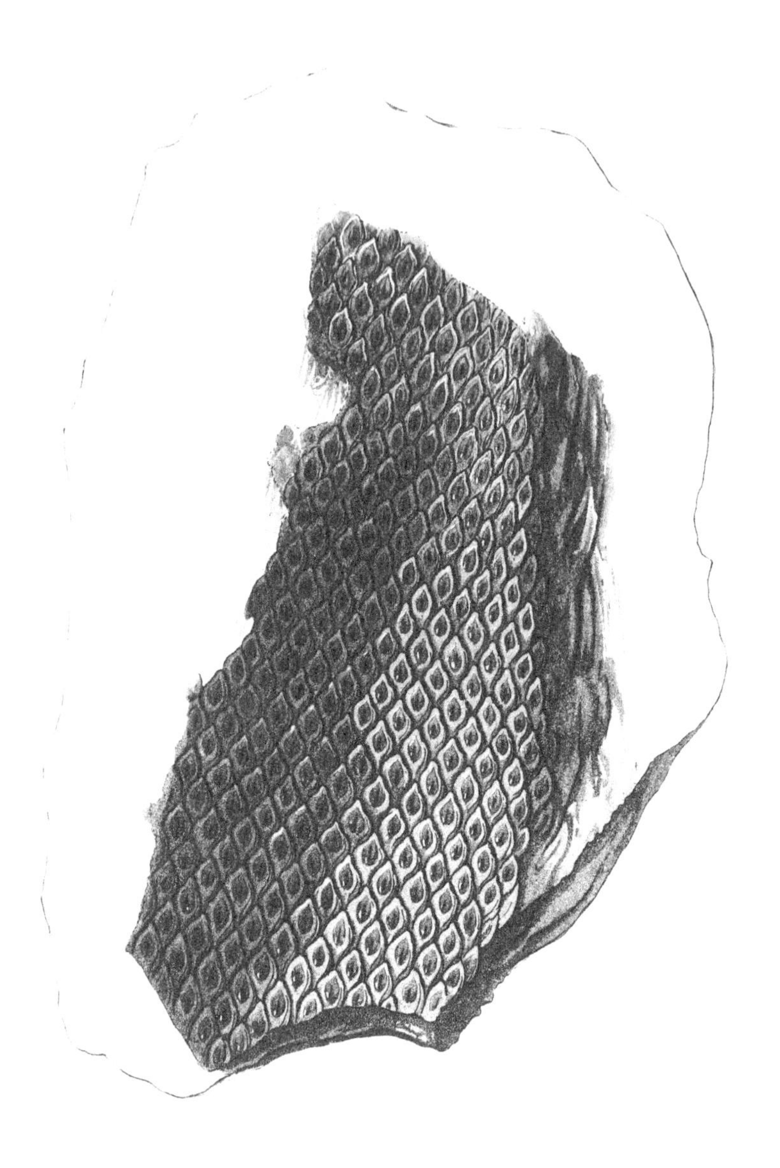

Pub by Messrs Ridgway London Jany 1837

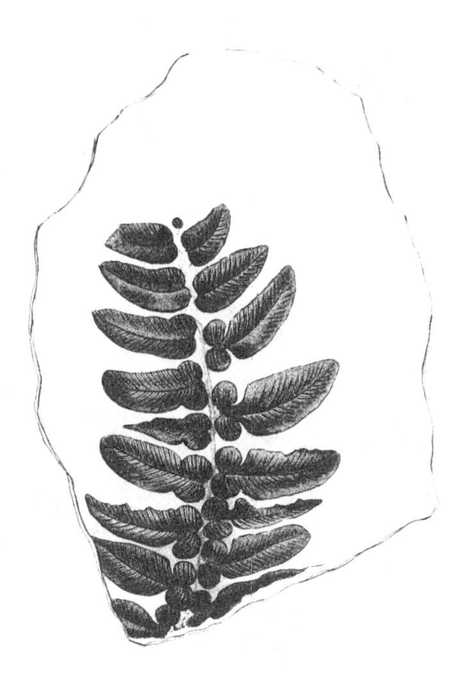

Pub by Messrs Ridgway London Jany 1837

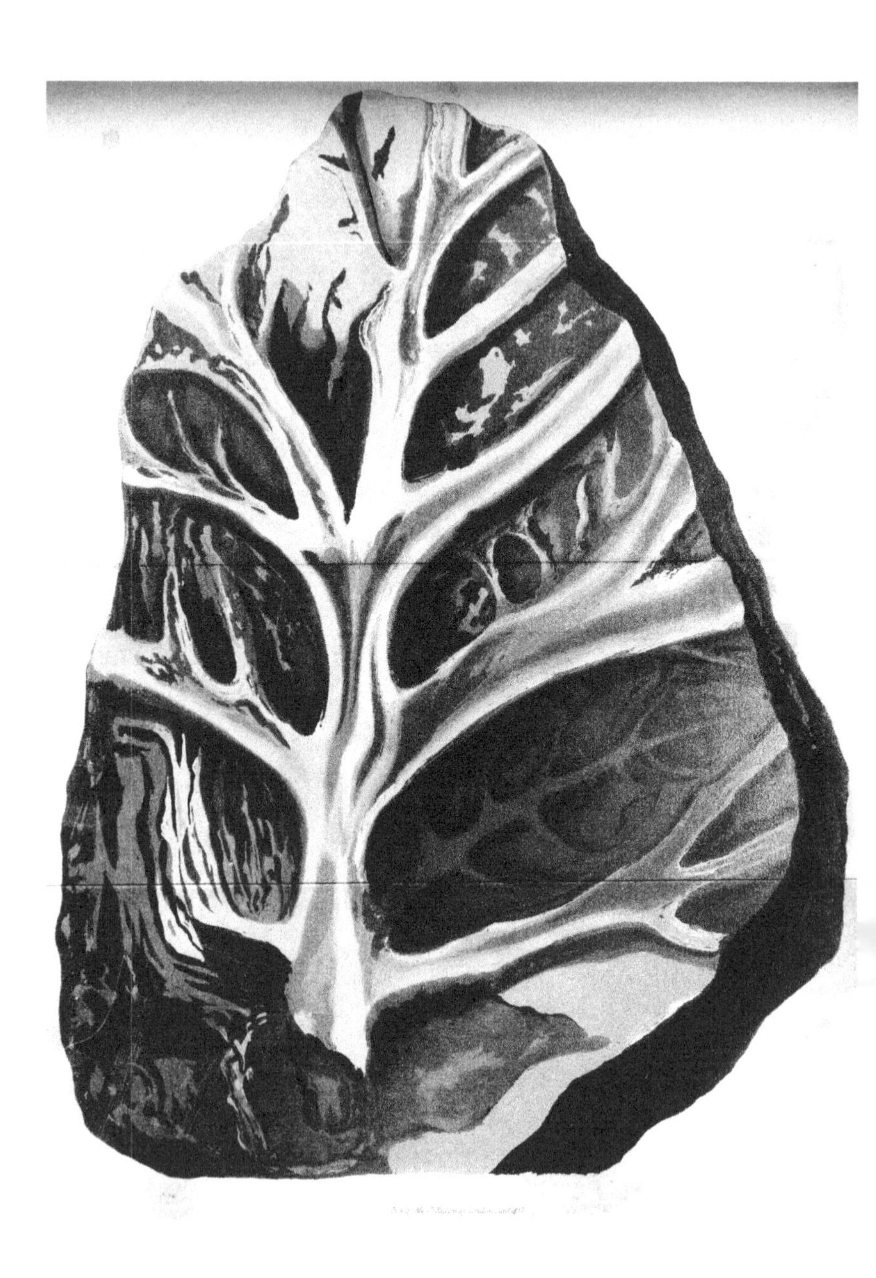

The material originally positioned here is too large for reproduction in this reissue. A PDF can be downloaded from the web address given on page iv of this book, by clicking on 'Resources Available'.

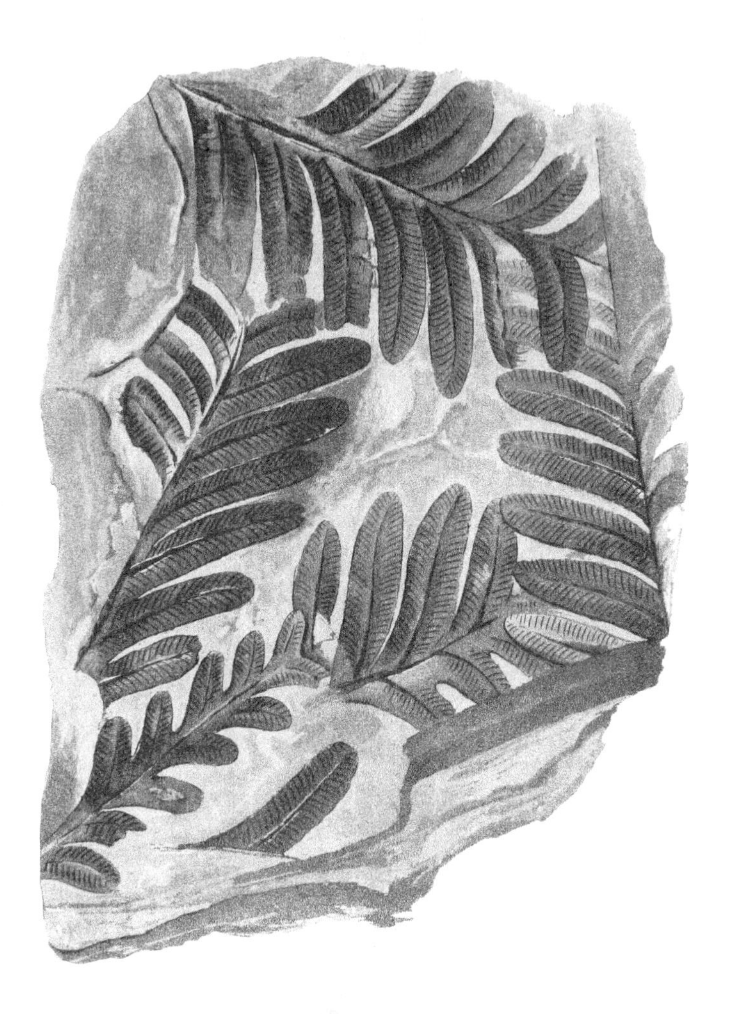

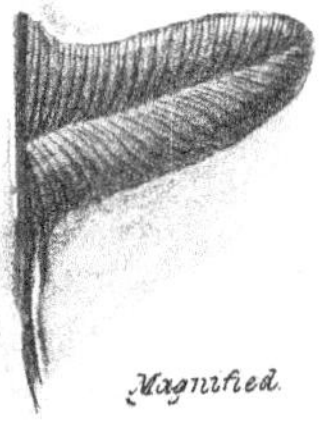

Magnified.

Pub. by Mess.rs Ridgway, London, Jan.y 1837

Pub. by Mess.rs Ridgway, London, Jan.y 1837

Magnified.

Pub. by Mefs.rs Ridgway, London. July 1837.

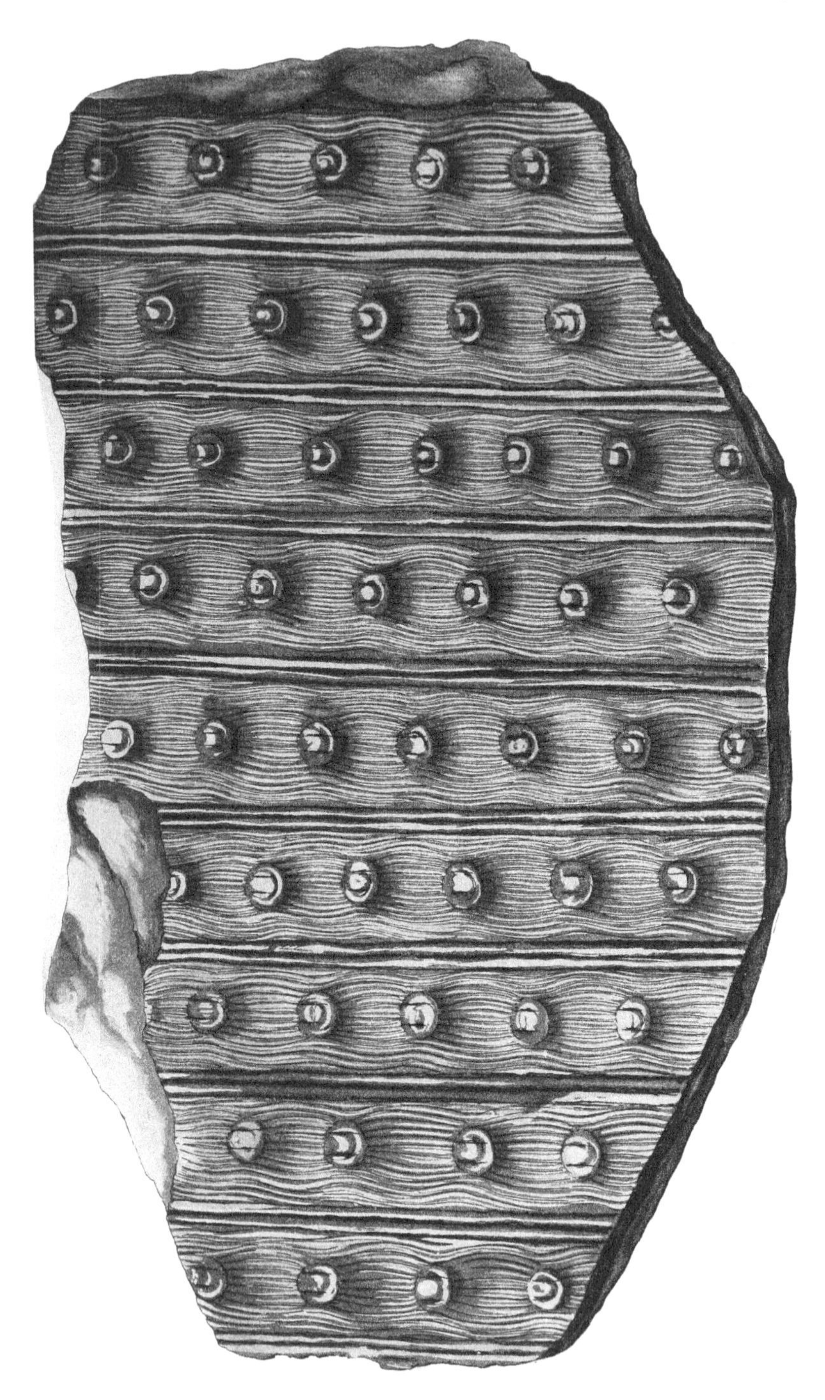

Pub: by Mefs.rs Ridgway, London, July 1837.

Pub by Messrs Ridgway, London, July 1837.

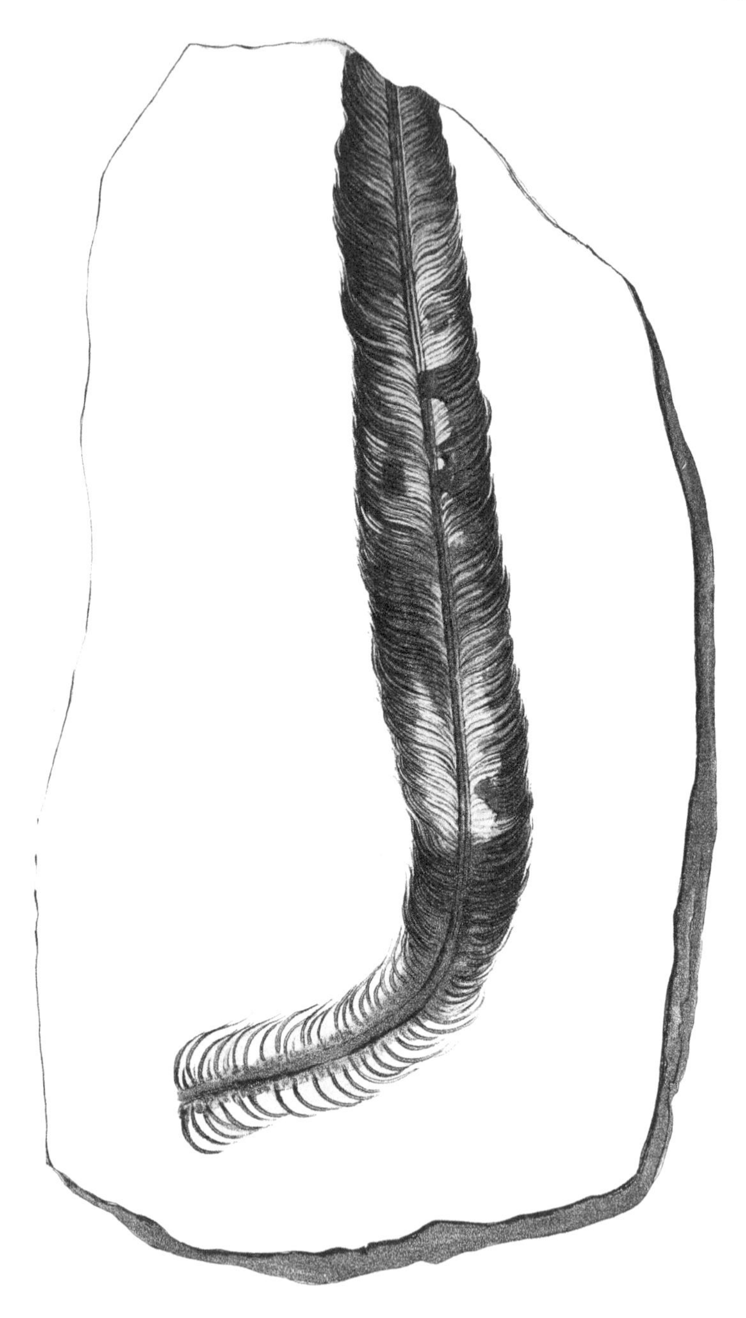

Pub: by Mefs.rs Ridgway, London, July 1837.

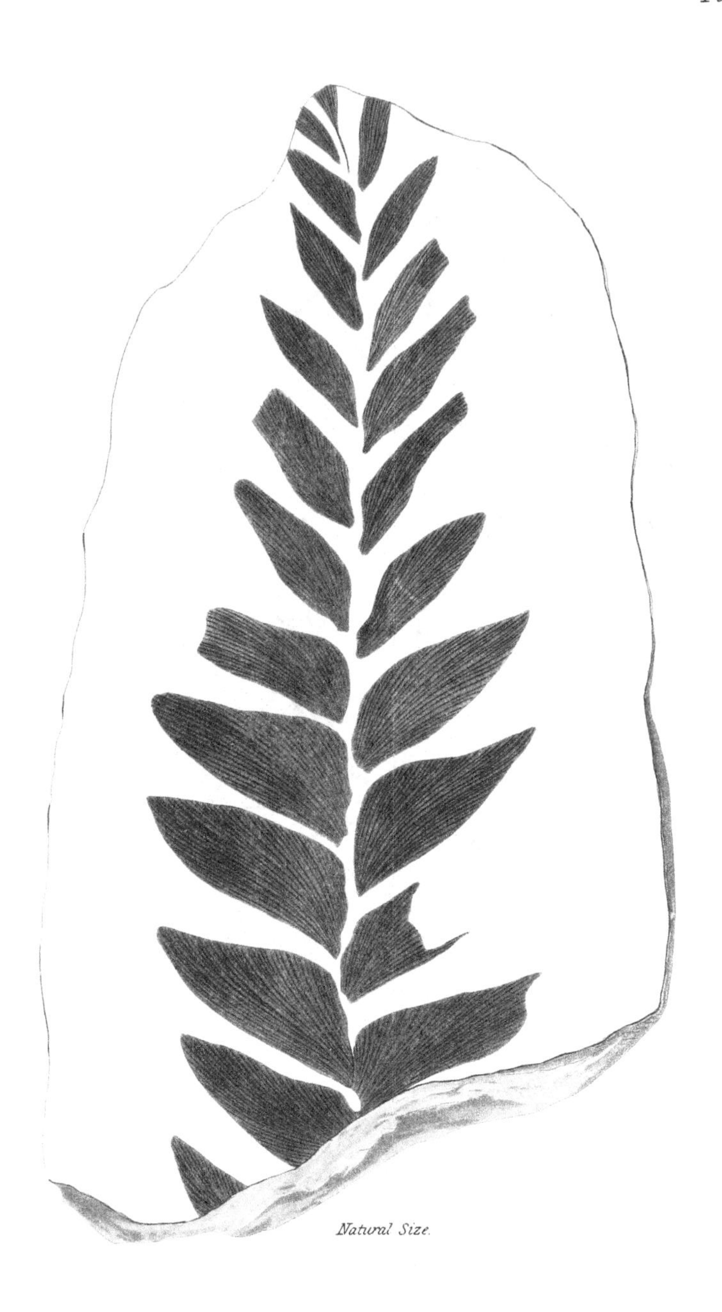

Natural Size.

Pub. by Mefs.rs Ridgway, London July, 1837.

Natural Size.

Pub. by Mefs.rs Ridgway, London, July 1837.

A

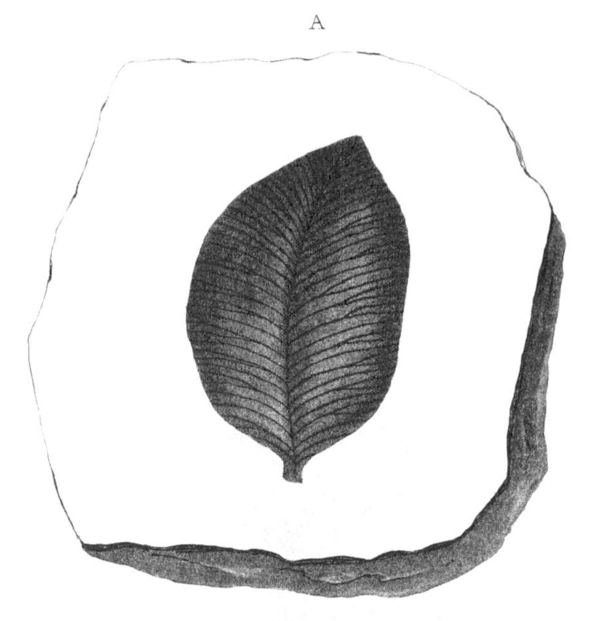

Natural Sizè.

B

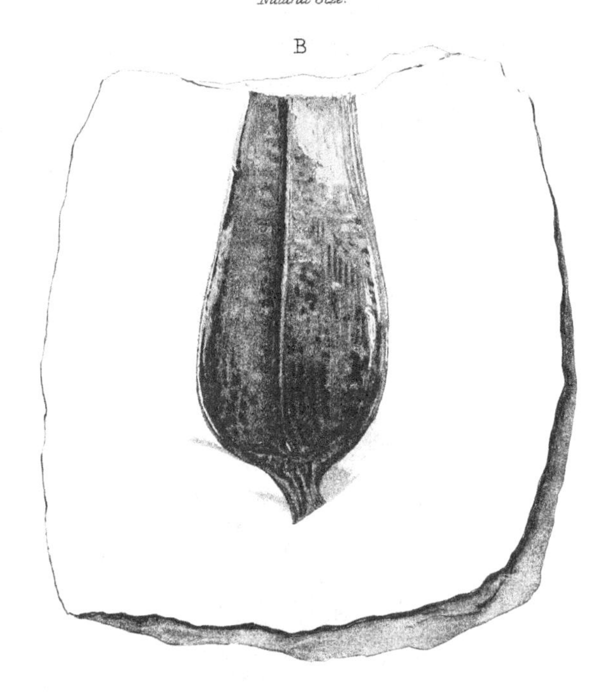

Pub. by Meſs.rs Ridgway, London July 1837.

Pub. by Mefs.rs Ridgway, London, July 1837.

Pub. by Messrs Ridgway, London, July, 1837.

The material originally positioned here is too large for reproduction in this reissue. A PDF can be downloaded from the web address given on page iv of this book, by clicking on 'Resources Available'.

Pub. by Mefs.rs Ridgway, London, July 1837.

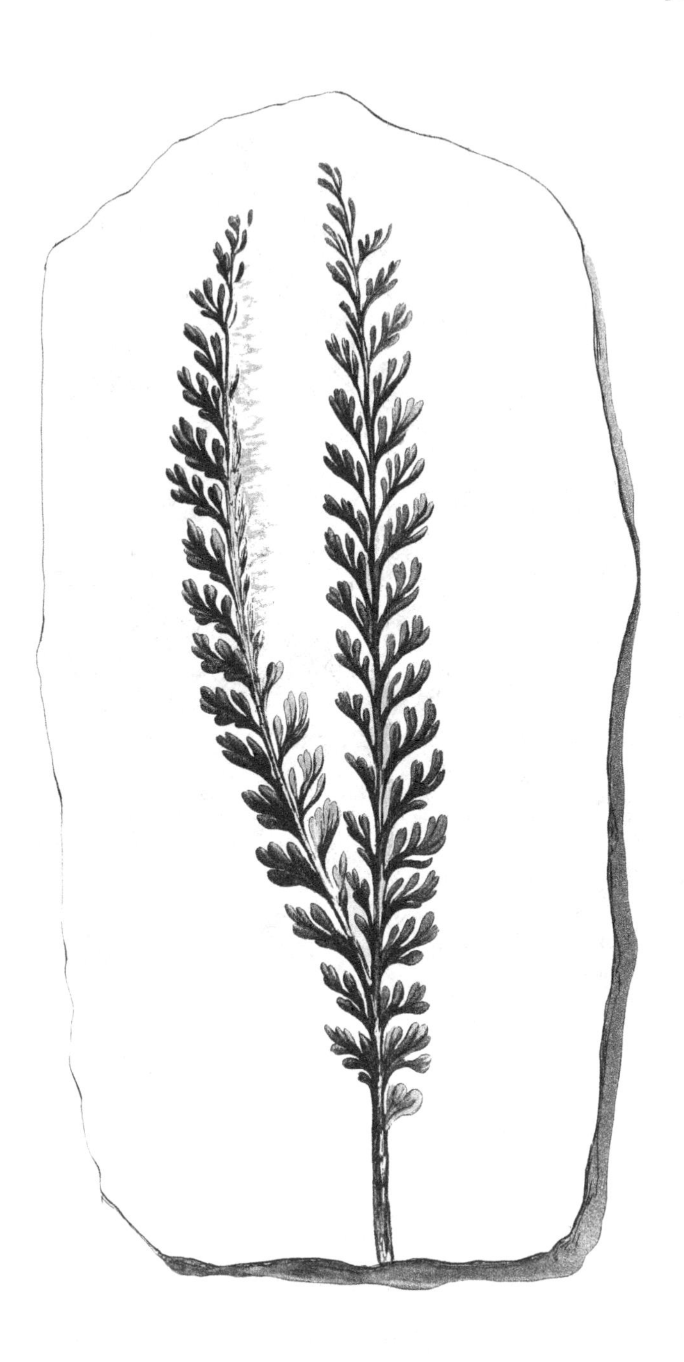

Pub. by Meſs.rs Ridgway, London, July 1837.

The material originally positioned here is too large for reproduction in this reissue. A PDF can be downloaded from the web address given on page iv of this book, by clicking on 'Resources Available'.

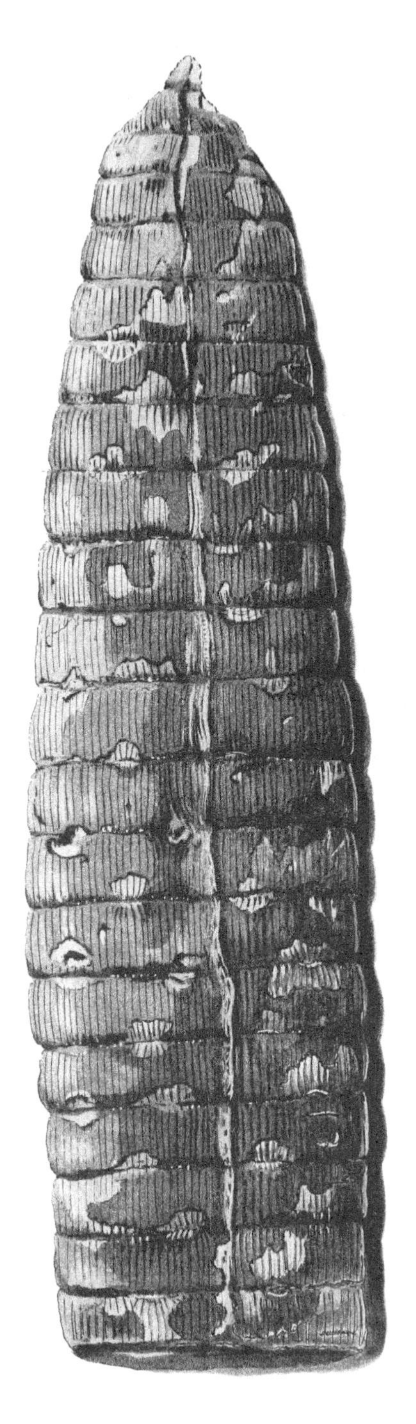

Pub. by Mess.rs Ridgway London July 1837.

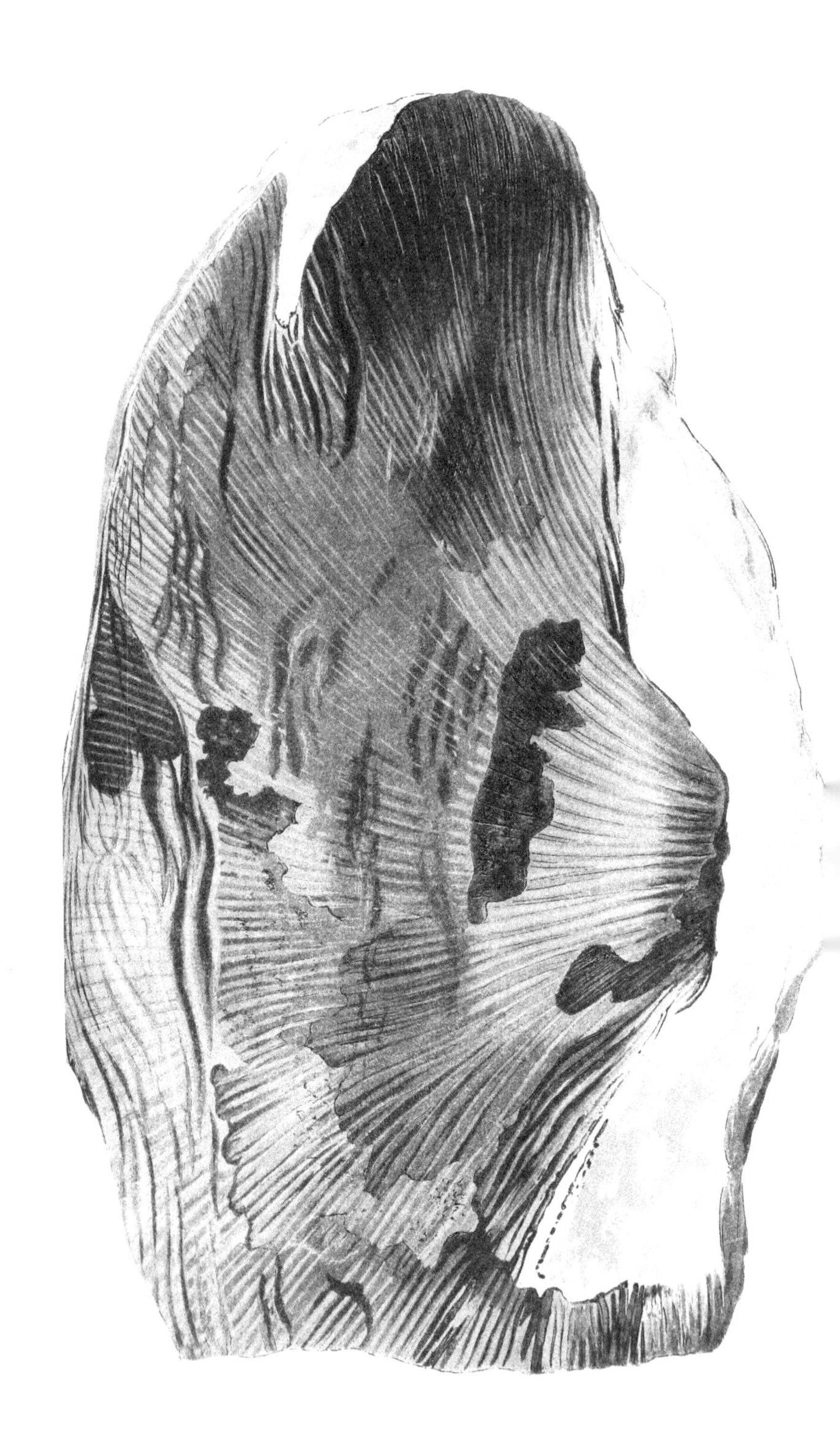

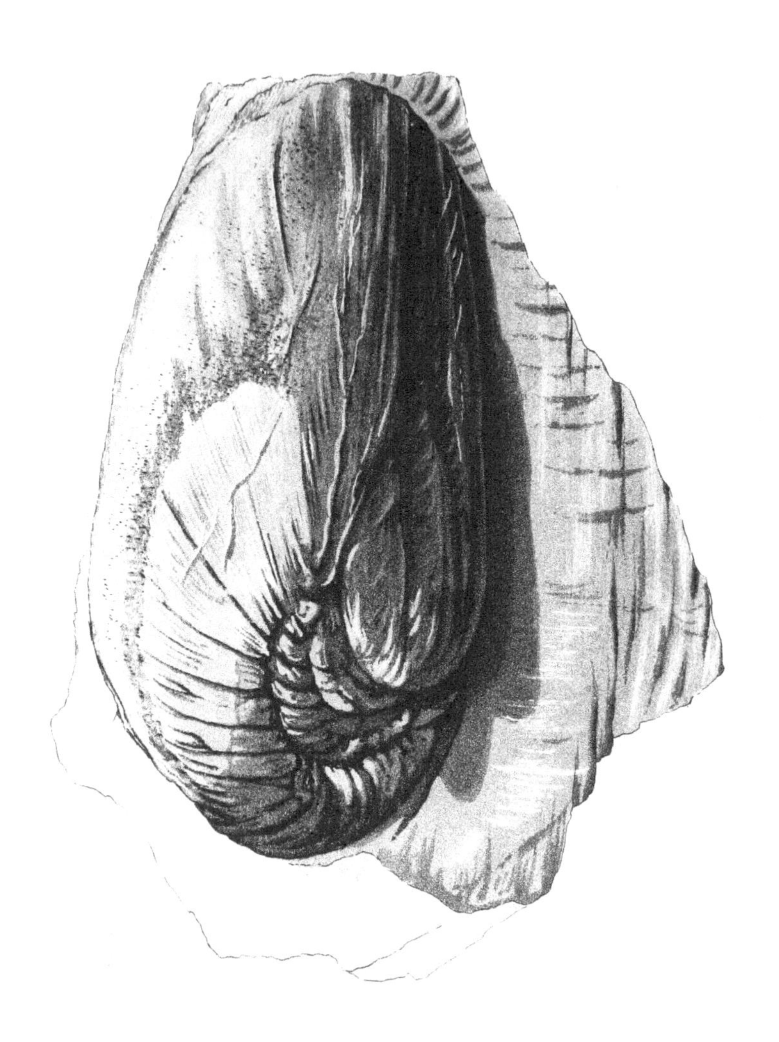

Pub. by Mefs.rs Ridgway. London. July 1837.

Plat

Pub: by Mess.rs Ridgway, London, July 1837.

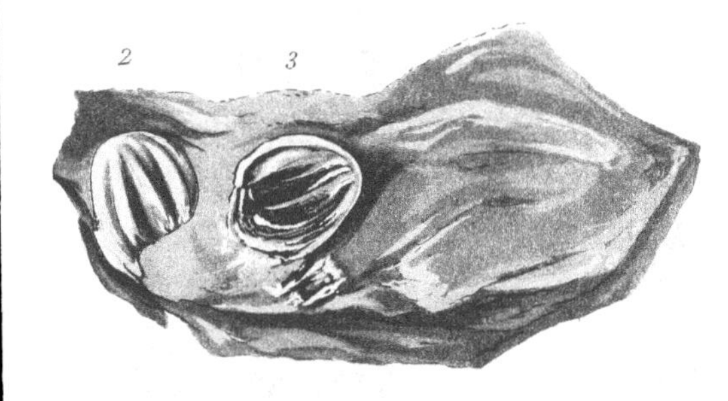

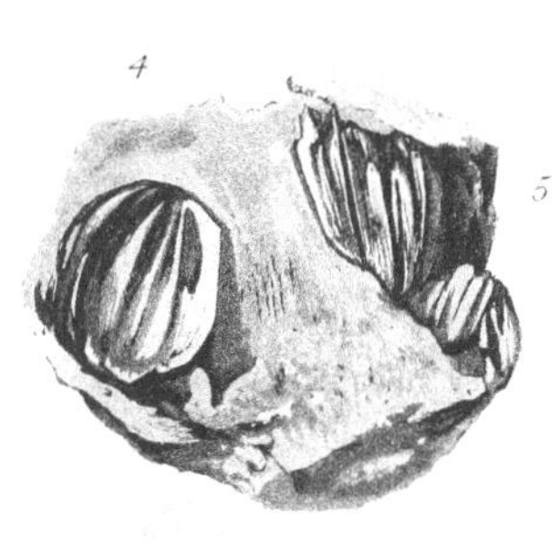

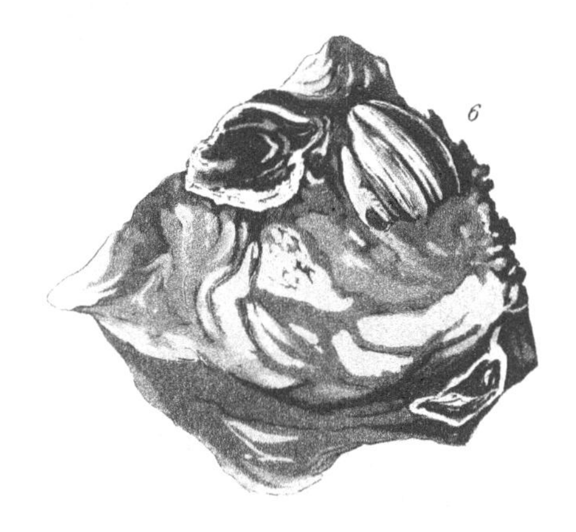

Pub. by Mefs.rs Ridgway, London July 1837.

Pub. by Mefs.rs Ridgway, London, July 1837.

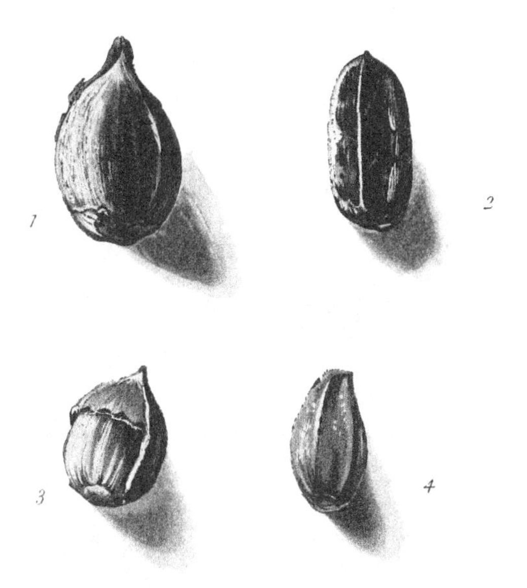

Pub: by Mefs.rs Ridgway, London, July 1837.

Magnified.

Pub: by Mefs.rs Ridgway, London, July 1837.

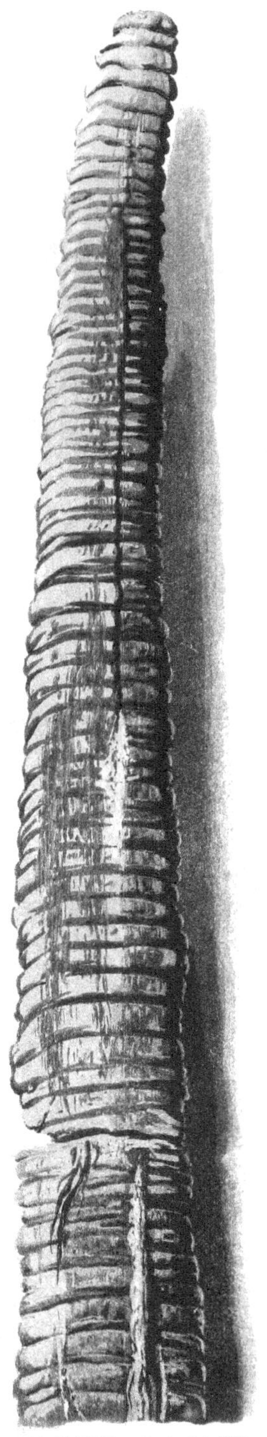

Pub. by Messrs. Ridgway, London, July 1837.

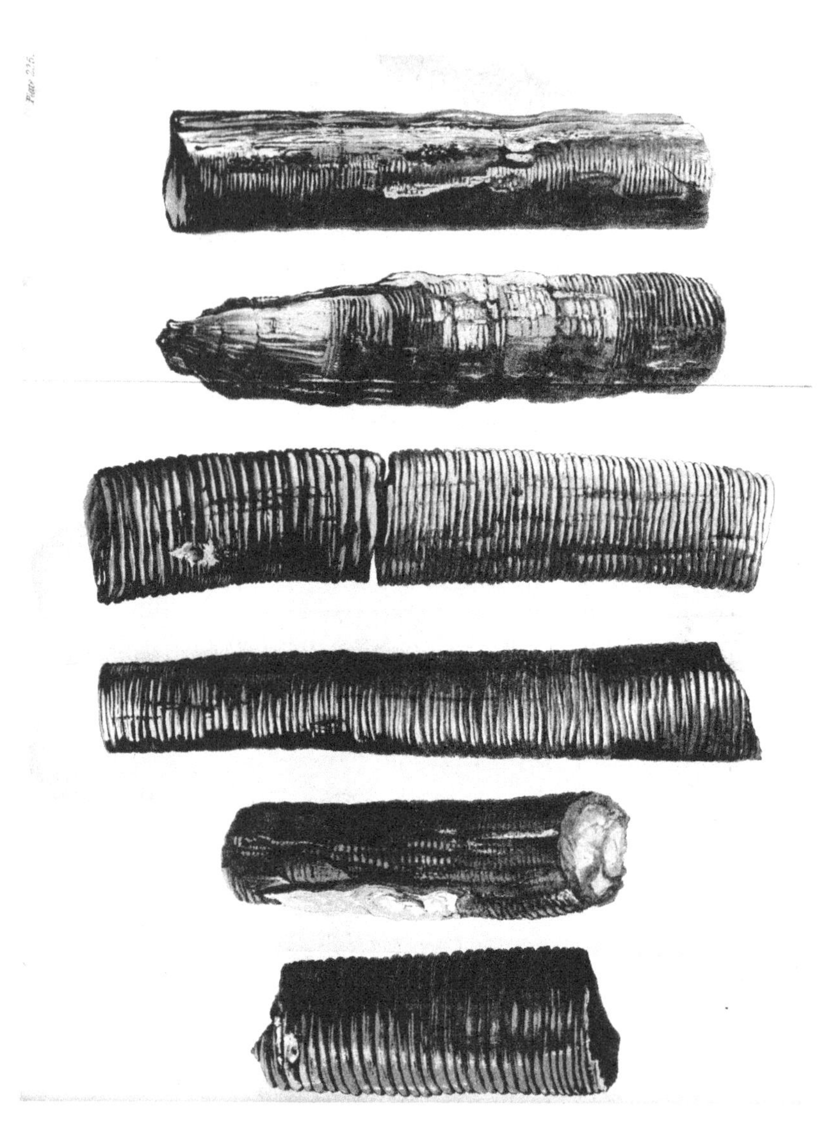

The material originally positioned here is too large for reproduction in this reissue. A PDF can be downloaded from the web address given on page iv of this book, by clicking on 'Resources Available'.

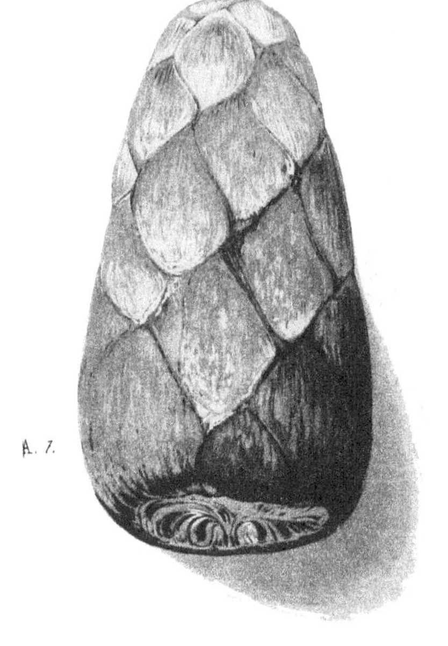

A. 1.

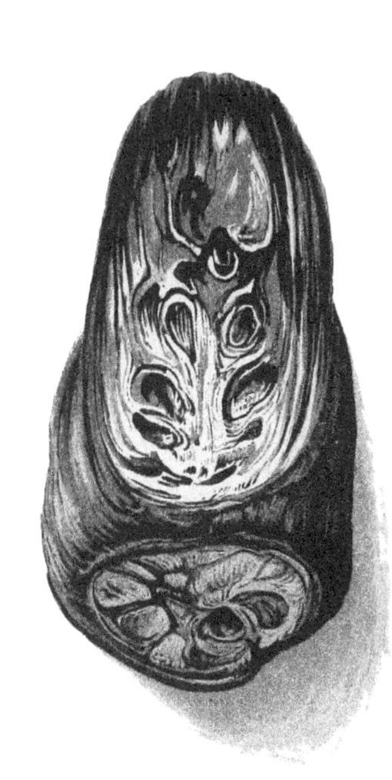

A. 2

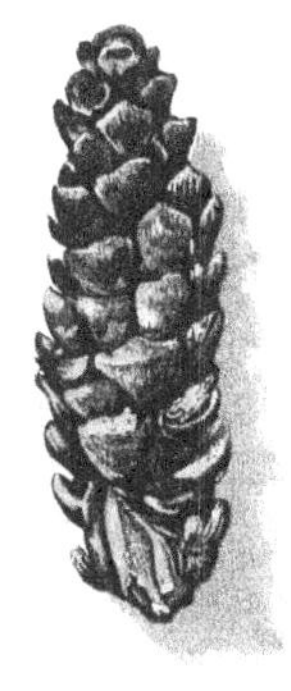

B. 1.

B. 2.

Pub by Mefs.rs Ridgway, London, July 1837.

Plate 2

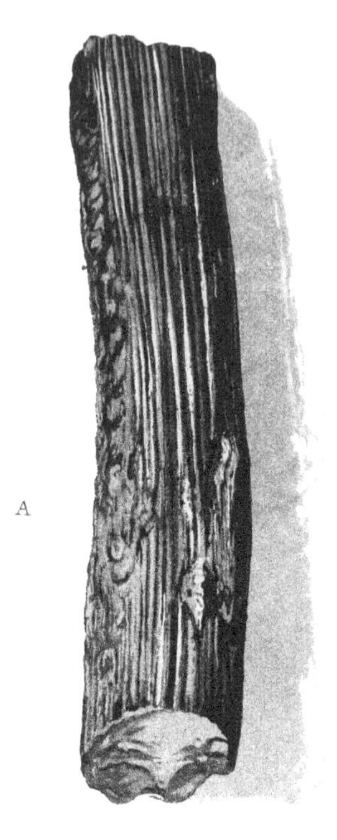

Pub by Mess.rs Ridgway, London, July 1837.

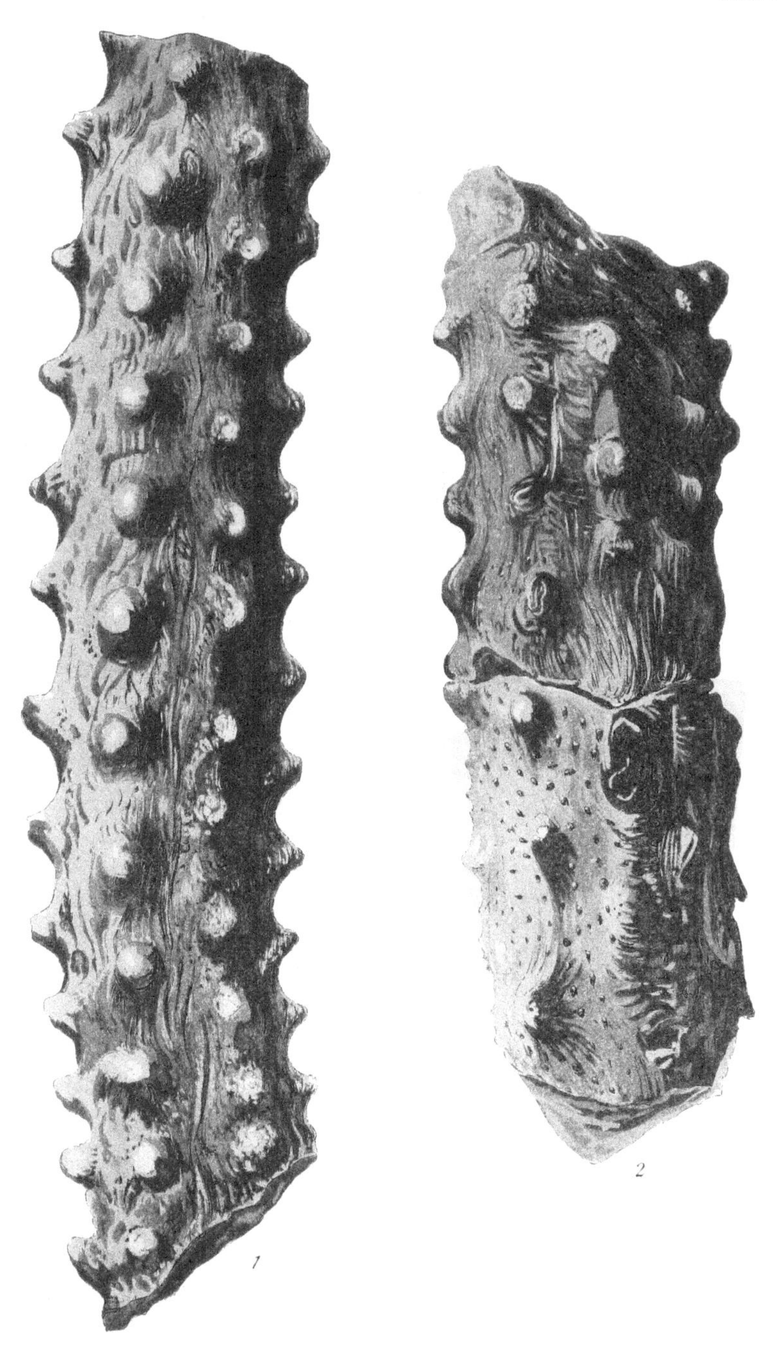

Pub: by Mess.rs Ridgway. London. July 1837.

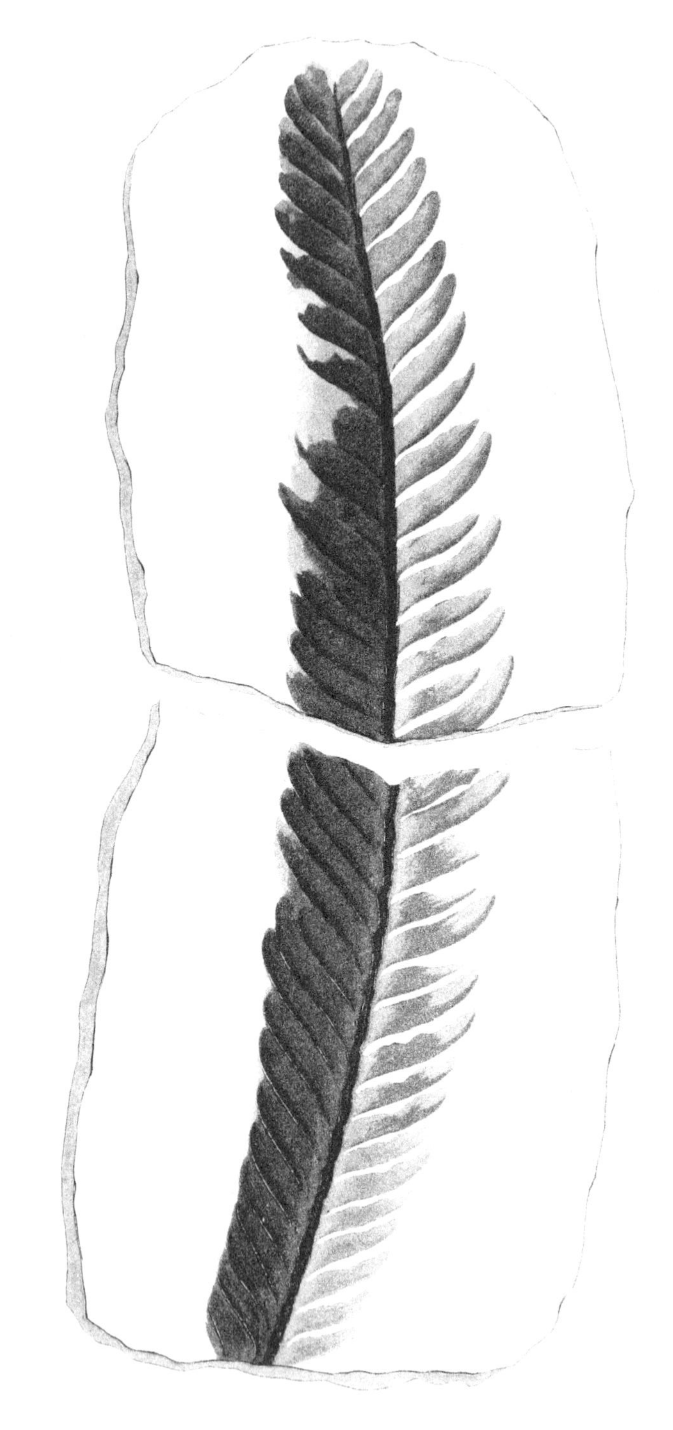

Pub. by Mess.rs Ridgway. London. July 1837.

Magnified

Pub: by Messrs Ridgway, London, July 1837.

Printed in the United States
By Bookmasters